Edible Fungi: Processing, Storage Preservation, Disease Control, and Potential Bioactivities

Edible Fungi: Processing, Storage Preservation, Disease Control, and Potential Bioactivities

Guest Editors

Demei Meng
Fansheng Cheng

MDPI

Basel • Beijing • Wuhan • Barcelona • Belgrade • Novi Sad • Cluj • Manchester

Guest Editors
Demei Meng
College of Food Science
and Engineering
Tianjin University of
Science & Technology
Tianjin
China

Fansheng Cheng
College of Food Science
and Engineering
Qingdao Agricultural University
Qingdao
China

Editorial Office
MDPI AG
Grosspeteranlage 5
4052 Basel, Switzerland

This is a reprint of the Special Issue, published open access by the journal *Foods* (ISSN 2304-8158), freely accessible at: https://www.mdpi.com/journal/foods/special_issues/204JQTOF5C.

For citation purposes, cite each article independently as indicated on the article page online and as indicated below:

Lastname, A.A.; Lastname, B.B. Article Title. *Journal Name* **Year**, *Volume Number*, Page Range.

ISBN 978-3-7258-5981-8 (Hbk)
ISBN 978-3-7258-5982-5 (PDF)
https://doi.org/10.3390/books978-3-7258-5982-5

Contents

About the Editors

Demei Meng

Demei Meng, an Associate Professor at Tianjin University of Science and Technology, in Tianjin, China, has focused her research career on postharvest quality preservation and disease control in edible fungi and horticultural produce. Her work centers on elucidating the physiological and molecular mechanisms that govern postharvest changes. She has led 19 research projects, including those supported by key national programs such as the National Key R&D Program and the National Natural Science Foundation of China. Her scholarly contributions are evidenced by nearly 100 publications, including 30 papers in high-ranking international journals, an ESI Highly Cited Paper, and over 1900 citations, yielding an H-index of 22.

Fansheng Cheng

Fansheng Cheng is a Professor of food science and engineering at the Qingdao Agricultural University, Qingdao China, where he teaches Food Biotechnology and Fruit and Vegetable Storage and Processing. He is also an edible mushroom industry consultant member in modern agriculture systems. He is dedicated to utilizing green preservation technology for horticulture produce, with a particular focus on mushroom postharvest biology, preservation technology and processing technology. He has expertise in molecular biology, as well as multiomics analysis in edible mushroom studies. Additionally, he also conducts experimental studies in fermentation technology using fungi to produce industrial enzymes.

foods MDPI

Editorial

Edible Fungi: Processing, Storage Preservation, Disease Control, and Potential Bioactivities

Demei Meng [1,*] and Fansheng Cheng [2,*]

1 College of Food Science and Engineering, Tianjin University of Science & Technology, Tianjin 300457, China
2 College of Food Science and Engineering, Qingdao Agricultural University, Qingdao 266109, China
* Correspondence: mengdm@tust.edu.cn (D.M.); fscheng@qau.edu.cn (F.C.); Tel.: +86-022-60912479 (D.M.); +86-0532-58957771 (F.C.)

Edible mushrooms have long been recognized for their nutritional and health benefits. However, their high moisture content and active metabolism render them highly perishable post-harvest, resulting in accelerated quality deterioration, including cap browning, stipe elongation, texture softening, and microbial spoilage [1]. These issues pose significant challenges for storage, transportation, and marketability, resulting in substantial economic losses [2]. Consequently, the development of effective preservation and disease control strategies is of the utmost importance for the mushroom industry.

Concurrently, mounting scientific evidence highlights the presence of diverse bioactive compounds in mushrooms, including polysaccharides, terpenoids, and phenolic compounds, which contribute to health-promoting properties such as antioxidant, immunomodulatory, and anti-cancer effects [3,4]. The utilization of these compounds in the development of functional foods, nutraceuticals, and even pharmaceuticals constitutes a pivotal strategy for enhancing the economic value of mushroom industry [5].

This editorial synthesizes recent advances and outlines future perspectives for addressing the challenges and opportunities in the edible mushroom sector, emphasizing sustainable practices and value addition.

This Special Issue, entitled "Edible Fungi: Processing, Storage Preservation, Disease Control, and Potential Bioactivities," successfully showcases the latest research advances in the field. The eight research papers included in this issue explore key challenges and innovative solutions in the edible mushroom industry from multiple perspectives.

In the area of bioactive component research, Adamczyk et al. conducted an in-depth structural characterization of (1→3)-α-D-glucan from Pleurotus djamor, revealing that this polysaccharide with a molecular weight of 552 kDa contains 86.4% (1→3) linked glucosyl units, laying a theoretical foundation for developing new functional materials.

In the field of nutritional enhancement applications, Hsu et al., through a human clinical trial, confirmed that pulsed UV-treated Pleurotus citrinopileatus significantly increased serum vitamin D_2 levels. The high-dose group (100 g/day) showed a more than 10-fold increase in vitamin D_2 levels, while concomitantly reducing parathyroid hormone levels by 37.6%, providing an effective mushroom-based solution for addressing vitamin D deficiency.

Regarding disease control technologies, Lei Zhang et al. found that carvacrol effectively inhibits Pseudomonas tolaasii. Its mechanism of action includes the disruption of bacterial cell membranes, the activation of the mushroom's defense system, and the promotion of the accumulation of antimicrobial substances. The efficacy of low-concentration fumigation at 20 μmol/L in the management of brown blotch disease was demonstrated.

In the area of preservation technology optimization, Yuxian Yang et al. demonstrated that controlled atmosphere storage with 1–3% O_2 and 15–17% CO_2 effectively delayed browning in Agaricus bisporus. In a separate study, Yalong Guo et al. optimized the best parameters for cold plasma treatment (95 kV, 130 Hz, 10 min) using response surface methodology, significantly extending mushroom shelf life.

In the field of pathogen genomics, Yufei Lan et al. made a significant contribution by completing the genome sequencing of three cobweb disease pathogenic fungi for the first time, revealing characteristics of their pathogenicity-related genes, providing a molecular basis for disease control.

In terms of processing and utilization innovation, Wenliang Wang et al. discovered that the incorporation of 10% Flammulina velutipes soluble dietary fiber improved noodle quality. Concurrently, the review by Akruti Singh et al. systematically summarized the bioactive components and health benefits of 11 edible mushroom species.

These research findings provide important theoretical support and technical foundations for the sustainable development of the edible mushroom industry, thereby promoting the transition from fundamental research to its practical application in industry. Future research will persist in concentrating on green control, precise preservation, and high-value utilization, thereby further unleashing the development potential of the edible mushroom industry.

Prospective Research Directions

Despite significant progress, future research should focus on several key areas to overcome existing challenges and fully exploit the potential of edible mushrooms:

Green and Safe Disease Control Technologies: Developing strategies based on plant essential oils (e.g., thymol), antagonistic microorganisms, and induced resistance to reduce reliance on chemical pesticides, ensuring product safety and environmental sustainability.

Integrated Preservation Methods: Combining physical techniques (e.g., cold plasma, pulsed light, and edible coatings), biological control, and smart packaging to establish a full-chain quality assurance system from postharvest to consumption.

Efficient Comprehensive Utilization of Bioactive Compounds: Applying advanced extraction technologies (e.g., ultrasound-assisted, microwave-assisted, enzymatic, and supercritical fluid extraction) to obtain bioactive compounds. Research should focus on elucidating their structure–activity relationships and developing high-value products, such as functional foods, nutraceuticals, and cosmeceuticals, to address product homogenization and drive industry upgrading.

The mushroom industry is evolving into an independent agricultural system, promising to form a tripartite agricultural (plant, animal, and mushroom agriculture). As a high-quality protein source with an amino acid profile comparable to animal proteins, mushrooms can alleviate protein supply pressures, particularly in developing countries [4]. Market trends, such as a 223-fold consumption increase in Japan over two decades, reflect growing consumer awareness and a solid industrial foundation. Supported by favorable policies, the utilization of abundant agricultural waste, and easily transferable cultivation techniques, the mushroom industry is poised to become a powerhouse for rural economic development worldwide. It creates new income streams, diversifies agricultural production, and provides new momentum for sustainable agricultural transformation, particularly in developing regions.

Author Contributions: Writing—original draft preparation, D.M.; writing—reveiw and editing, F.S. All authors have read and agreed to the published version of the manuscript.

Conflicts of Interest: The authors declare no conflicts of interest.

List of Contributions:

1. Adamczyk, P.; Komaniecka, I.; Siwulski, M; Wlizło, K.; Junka, A.; Nowak, A.; Kowalczyk, D.; Waśko, A.; Lisiecka, J.; Grzymajło, M.; et al. (1→3)-α-d-Glucan is an important component of the cell wall of most fungi. *Foods* **2025**, *14*, 1272. https://doi.org/10.3390/foods14071272
2. Hsu, C.-C.; Chen, C.-M.; Ju, Y.-M.; Wu, Y.-C.; Hsieh, H.-M.; Yang, S.-H.; Su, C.-T.; Fang, T.-C.; Setyaningsih, W.; Li, S.-C. Effects of Consuming Pulsed UV Light-Treated Pleurotus citrinopileatus on Vitamin D Nutritional Status in Healthy Adults. *Foods* **2025**, *14*, 259. https://doi.org/10.3390/foods14020259
3. Zhang, L.; Song, R.; Shi, Z.; Yuan, S.; Jiao, L.; Ma, M.; Wang, X.; Chen, L.; Liu, X.; Meng, D. Carvacrol Effectively Inhibits Pseudomonas tolaasii In Vitro and Induces Resistance to Brown Blotch Disease in Postharvest Agaricus bisporus. *Foods* **2024**, *13*, 3689. https://doi.org/10.3390/foods13223689
4. Yang, Y.; Jia, O.; Li, Y.; Feng, B.; Chang, M.; Meng, J.; Deng, B. Effect of high CO_2 controlled atmosphere storage on Postharvest Quality of Button mushroom (*Agaricus bisporus*). *Foods* **2024**, *13*, 3486. https://doi.org/10.3390/foods13213486
5. Guo, Y.; Xia, S.; Shi, C.; Ma, N.; Pei, F.; Yang, W.; Hu, Q.; Kimatu, B.M.; Fang, D. The Effect of Cold Plasma Treatment on the Storage Stability of Mushrooms (*Agaricus bisporus*). *Foods* **2024**, *13*, 3393. https://doi.org/10.3390/foods13213393
6. Lan, Y.; Cong, Q.; Yu, Q.; Liu, L.; Cui, X.; Li, X.; Wang, Q.; Yang, S.; Yu, H.; Kong, Y. Genome Sequencing of Three Pathogenic Fungi Provides Insights into the Evolution and Pathogenic Mechanisms of the Cobweb Disease on Cultivated Mushrooms. *Foods* **2024**, *13*, 2779. https://doi.org/10.3390/foods13172779
7. Wang, W.; Yang, S.; Wang, L.; Hou, F.; Song, S.; Wang, Y.; Gong, Z.; Jia, F. Effect of Flammulina velutipes Soluble Dietary Fiber on Dough Processing Characteristics and Micro-Fermented Dried Noodles Quality Properties. *Foods* **2024**, *13*, 2764; https://doi.org/10.3390/foods13172764
8. Singh, A.; Saini, R.K.; Kumar, A.; Chawla, P.; Kaushik, R. Mushrooms as Nutritional Powerhouses: A Review of Their Bioactive Compounds, Health Benefits, and Value-Added Products. *Foods* **2025**, *14*, 741. https://doi.org/10.3390/foods14050741

References

1. Valverde, M.E.; Hernández-Pérez, T.; Paredes-López, O. Edible mushrooms: Improving human health and promoting quality life. *Int. J. Microbiol.* **2015**, *2015*, 376387. [CrossRef] [PubMed]
2. Liu, S.; Liu, H.; Li, J.; Wang, Y. Advance in Elements of Wild Edible Mushrooms. *J. Fungi* **2022**, *8*, 964. [CrossRef] [PubMed]
3. Rathore, H.; Prasad, S.; Sharma, S. Mushroom nutraceuticals for improved nutrition and better human health: A review. *PharmaNutrition* **2017**, *5*, 35–46. [CrossRef]
4. Friedman, M. Chemistry, Nutrition, and Health-Promoting Properties of Hericium erinaceus(Lion's Mane) Mushroom Fruiting Bodies and Mycelia and Their Bioactive Compounds. *J. Agric. Food Chem.* **2016**, *63*, 7108–7123. [CrossRef] [PubMed]
5. Wasser, S.P. Medicinal mushrooms in human clinical studies. Part I. Anticancer, oncoimmunological, and immunomodulatory activities. *Int. J. Med. Mushrooms* **2017**, *19*, 279–317. [CrossRef] [PubMed]

foods

MDPI

Article

Effect of *Flammulina velutipes* Soluble Dietary Fiber on Dough Processing Characteristics and Micro-Fermented Dried Noodles Quality Properties

Wenliang Wang [1,†], Shuang Yang [1,2,†], Lihong Wang [2], Furong Hou [1], Shasha Song [1], Yansheng Wang [1], Zhiqing Gong [1] and Fengjuan Jia [1,*]

1 Institute of Agro-Food Science and Technology, Shandong Academy of Agricultural Sciences, Jinan 250100, China; cywwl@163.com (W.W.)
2 Department of Life Science and Food Engineering, Hebei University of Engineering, Handan 056200, China
* Correspondence: jfj.5566@163.com
† These authors contributed equally to this work.

Abstract: Our research focused on the integration of *Flammulina velutipes* soluble dietary fiber (Fv-SDF) into wheat flour during the production of dried noodles, delving into the impact of different addition ratios of Fv-SDF on both dough processing characteristics and the quality of the micro-fermented dried noodles. The viscometric and thermodynamic analyses revealed that Fv-SDF notably improved the thermal stability of the mix powder, reduced viscosity, and delayed starch aging. Additionally, Fv-SDF elevated the gelatinization temperature and enthalpy value of the blend. Farinograph Properties and dynamic rheology properties further indicated that Fv-SDF improved dough formation time, stability time, powder quality index, and viscoelasticity. Notably, at a 10% Fv-SDF addition, the noodles achieved the highest sensory score (92) and water absorption rate (148%), while maintaining a lower dry matter loss rate (5.2%) and optimal cooking time (142 s). Gas chromatography-ion mobility spectrometry (GC-IMS) analysis showed that 67 volatile substances were detected, and the contents of furfural, 1-hydroxy-2-acetone, propionic acid, and 3-methylbutyraldehyde were higher in the Fv-SDF 10% group. These 10% Fv-SDF micro-fermented noodles were not only nutritionally enhanced, but also had a unique flavor. This study provides a valuable theoretical basis for the industrial application of *F. velutipes* and the development of high-quality dried noodles rich in Fv-SDF.

Keywords: soluble dietary fiber; micro-fermented dried noodles; farinograph properties; dynamic rheology properties; nutrition; flavor

1. Introduction

Flammulina velutipes is known as enoki mushroom, golden mushroom, belonging to Basidiomycotina, Agaricales, Tricholomataceae, and Flammulina [1]. It is cultivated at large scales in East Asia, especially China, Japan, Vietnam and Korea [2]. *F. velutipes* is also one of the major edible mushrooms employed in factory cultivation, and it has medicinal, edible, and ornamental value, with immeasurable commercial potential [3]. *F. velutipes* contains many healthy nutrients, including dietary fiber, vitamins, minerals, organic acids, and other biologically active components [4]. The contents of polysaccharides and peptides in *F. velutipes* are 7% and 15%, respectively [5]. There is a wide range of biological as well as pharmacological activities associated with these compounds, such as antifungal, anticancer, anti-tumor, anti-inflammatory, and antioxidant properties [6]. Dietary fiber (DF) is one of the main nutritional components of *F. velutipes*, accounting for 32% of dry *F. velutipes* body [7]. Compared with the polysaccharides and peptides in *F. velutipes*, the rich content of DF is rarely mentioned and has not been effectively utilized.

The term DF refers to indigestible carbohydrates and lignin, which could be categorized as plant dietary fiber, animal dietary fiber, algae dietary fiber, or microbial dietary fiber [8]. According to the amount of water they can saturate, DF is commonly classified as insoluble (IDF) or soluble (SDF) [9]. IDF is principally composed of cellulose, lignin, and insoluble hemicellulose, while SDF is composed of soluble hemicellulose, pectin, gum, and oligosaccharides [10]. DF is a nonnutritive component of food, and several health benefits have been associated with DF. Studies have shown that DF can effectively inhibit the absorption of cholesterol in the digestive tract, prevent it from entering the bloodstream, and effectively prevent cardiovascular and cerebrovascular diseases caused by arteriosclerosis [11]. DF has good water and fat solubility and can combine with sugars and oils in the intestine to reduce the absorption of sugars and oils in the intestine [12]. It has been suggested that DF consumption could contribute to diabetes prevention by increasing satiety, reducing nutrient absorption, and decreasing weight [13]. In addition, DF also has a certain role in reducing the incidence of colon cancer, and breast cancer [14]. On the other hand, SDF has a broad application prospect as a food additive, stabilizer, gelling agent, and thickener in food processing based on its advantages of water retention, oil retention, and structural characteristics [10]. Moreover, the addition of SDF can extend the shelf life of fat-rich foods by increasing the antioxidant capacity of the emulsion [15]. The SDF of *F. velutipes* (Fv-SDF) is mainly composed of glucose, hemicellulose, dextran, pectin, oligosaccharides, etc, and the application of Fv-SDF in food processing has been rarely studied.

In the past few decades, people have been blindly pursuing fine grains, and many nutrients of traditional wheat flour have been seriously lost under various fine processing, especially dietary fiber [16]. The loss of nutrients is directly related to the eating quality and nutrition of flour products. SDF has high gelation, can be used as an emulsifier, and is easy to combine with food systems [17]. SDF is filled between the protein matrix and starch granules in the form of a binder. Although DF content in wheat flour is only 10–15%, it greatly influences the quality and taste of flour products [18]. Dried noodles are the staple food of many Asian countries and are one of the most produced and consumed flour products in China [19]. Considering the deep processing of grain and the sub-health of people, the importance of DF in grain processing and diet is highlighted [20]. At present, most of the research focuses on the influence of grain DF on the quality characteristics of ordinary dried noodles, while research on the influence of fruit and vegetable DF on the quality of micro-fermented dried noodles is rare. In recent years, micro-fermented dried noodles are one of the most popular high-quality dried noodles in the market, with a smooth appearance without depressions, and many fine holes in the internal structure [21]. The main advantages of micro-fermented dried noodles are enhanced nutritional value, improved digestion and absorption, rich taste, and easy digestion. Micro-fermentation technology, with its unique production technology, not only retains the advantages of traditional noodles but also significantly improves the flavor and texture of fermentation. Micro-fermented dried noodles have great development potential in the dried noodles market because of their good taste and suitability for both young and old people. However, no studies have been done to improve the quality of dried noodles by both adding Fv-SDF and applying micro-fermentation technology.

In this study, the effects of different doses of Fv-SDF on dough processing characteristics and the cooking characteristics of micro-fermented dried noodles were investigated. The application of Fv-SDF in food processing extended the edible mushroom industry chain and promoted the development of the edible mushroom industry, which was of great significance for the development of the edible mushroom processing industry. Fv-SDF micro-fermented dried noodles not only satisfied consumers' demand for nutritionally fortified foods but also enriched the varieties of dried noodles, which laid a foundation for developing dried noodles with more nutritional and health care functions and provided a reference for the application of edible fungi in food.

2. Materials and Methods

2.1. Preparation of Mixed Flour Dough and Fv-SDF Micro-Fermented Dried Noodles

The Fv-SDF was extracted from *F. valutipes* by ultrasound-assisted subcritical water extraction following the method of Yan et al. [22]. The noodles of Ginindza et al. [23] were adopted and improved. The mixed system of Fv-SDF and wheat flour (Jinlongyu Grains and Oils Food Co., Shanghai, China) was prepared by adding Fv-SDF to wheat flour at the proportions of 0%, 5.0%, 7.5%, 10.0%, 12.5%, and 15.0%, and mixing them evenly. Then, 1.0% salt, 1.2% yeast, and 35 mL distilled water were added in proportion. The mixture was mixed evenly, placed into a bread maker (EGBM010 bread machine, Electrolux Electric Co., Ltd., Shanghai, China), and stirred for 10 min. After the dough was mixed, it was kneaded by hand, and then put in the fermentation chamber (6D Fermentation chamber, Deer Ma Appliances, Foshan, China) for 9 min. The noodles were made with a noodle maker. The noodles were dried in a 40 °C blast drying oven (GZX-9240MBE Electric Blast Drying Oven, Shanghai, China) for 4 h to prepare Fv-SDF micro-fermented dried noodles.

2.2. Differential Scanning Calorimeter Analysis

DSC (DSC-250, TA Instruments-Waters, New Castle, NC, USA) was used to determine the thermodynamic properties of the mix powder [24]. The solution of the 4.0 mg sample (mix powder:water = 5 g:10 mL) was placed in a crucible and stored overnight at 4 °C to make the sample fully expand. During the measurement, the sample was heated from 25 °C to 100 °C at the speed of 10 °C/min. The samples were measured three times to obtain thermodynamic parameters. The thermodynamic parameters included initial gelatinization temperature (T_0/°C), peak temperature (Tp/°C), terminating gelatinization temperature (Tc/°C), and gelatinization enthalpy (ΔH, W/g).

2.3. Rapid Analysis of Viscosity

The gelatinization properties of the mixture were measured by RVA [25]. Then, 3 g of Fv-SDF complex and wheat flour were weighed and 25 mL of distilled water was added to make the total weight 28 g. Then, it was put into a rapid viscosity analyzer (RVA-Tec Master, Perten, Australia) for testing. The gelatinization properties included peak viscosity (cp) trough viscosity (cp), breakdown value (cp), final viscosity (cp), setback value (cp) peak time (min), and pasting temperature (°C).

2.4. Determination of Farinograph Properties

Farinograph (Micro-dough LAB, Sweden porton corp, Stockholm, Sweden) was used to measure the farinograph properties of the mixture, following the method of Zhang et al. [26] with slight modifications. The mixed system of Fv-SDF wheat flour was prepared by adding Fv-SDF to wheat flour at the ratios of 0%, 5.0%, 7.5%, 10.0%, 12.5%, and 15.0%. An amount of 50 g of the powder mixture was placed in the farinograph and the rotation speed was set at 61~65 r/min. The farinograph properties included consistency (FU), water absorption (%), dough development time (min), stability time (min), weakening degree (FU), and flour quality index (mm).

2.5. Determination of the Dynamic Rheology Properties

The dynamic rheology properties were measured according to Huang et al. [12,27] with slight modifications. Then, 4 g of Fv-SDF micro-fermented dough (mix powder:water = 100:40) was put in a rheometer (HR 20 Rheometer, TA Instruments, Waters Corporation, Milford, MA, USA) to determine its storage modulus (G'), loss modulus (G'') and loss tangent (tanδ). The frequency scanning was 0.1~500%, and the testing temperature was 25 °C.

2.6. Apparent Color Detection of Dried Noodles

The effect of the addition of Fv-SDF on the chromaticity value of micro-fermented dried noodles was measured by a portable precision colorimeter (CR-400, Konica Minolta, Tokyo, Japan) [28]. The CIE-Lab color space represents the color characteristics of noodles, with the L* value representing whiteness, the a* value representing red-greenness, and the b* value representing yellow-blue intensity. The probe of the colorimeter measured the samples of micro-fermented noodles and recorded the values of L*, a*, and b*. Each sample was tested 3 times, and the average value of the 3 tests was the final result.

2.7. Determination of the Best Cooking Time for Dried Noodles

The method of Noonim et al. [29] was used and improved. 10 g noodles were boiled in 500 mL of water for 1 min, and then a noodle was taken out every 15 s. The best cooking time for noodles is when the white starch core of the noodles disappears.

2.8. Determination of the Cooking Characteristics of Dried Noodles

The cooking characteristics of micro-fermented dried noodles were determined according to the method of Wang et al. [30] with slight modifications. 20 g of Fv-SDF micro-fermented dried noodles were put into 1 L of boiling water and cooked at the best cooking time. The water absorption and cooking loss rate of the noodles were calculated according to the following Formulas (1) and (2), respectively.

$$\text{Water absorption} = \frac{m_2 - m_1}{m_1} \times 100\% \quad (1)$$

$$\text{Cooking loss} = \frac{m_1 - m_3}{m_1} \times 100\% \quad (2)$$

where m_1 is the dry weight/g of Fv-SDF micro-fermented dried noodles; m_2 is the cooked mass/g of Fv-SDF micro-fermented dried noodles; m_3 is the dry weight/g of Fv-SDF micro-fermented dried noodles after cooking.

2.9. Determination of Sensory Evaluation of Dried Noodles

According to the method proposed by Yeoh et al. [31], the sensory evaluation of Fv-SDF micro-fermented dried noodles was determined. Six men and six women were selected for sensory evaluation of the noodles, and the scoring criteria are shown in Table S1. Twelve healthy panelists (6 females and 6 males, 20–28 years of age) were recruited from the Institute of Agro-Food Sciences and Technology, Shandong Academy of Agricultural Sciences (Jinan, China). Before the sensory test, these 12 members needed to undergo a week-long systematic sensory evaluation training according to the GB/T 16291.1-2012 standard [32]. All sensory tests were conducted in an air-conditioned room (22 °C) with separate compartments. We adhered to the ethical principles of sensory research at the Agro-Food Sciences and Technology, Shandong Academy of Agricultural Sciences. These principles were reviewed by the Research Ethics Committee at the Institute of Agro-Food Sciences and Technology, Shandong Academy of Agricultural Sciences (Statement 11/2023). Before participating in this study, each participant provided written informed consent.

2.10. Determination of Flavor Compounds of Dried Noodles

According to the descriptive analysis method of Hou et al. [32], the flavor substances of Fv-SDF micro-fermented dried noodles were determined. The volatile compounds were analyzed by chromatography ion mobility spectrometry (GC-IMS) instrument (Gesellschaftfür Analytische Sensorsysteme mbH (G.A.S.), Dortmund, Germany). An amount of 3 g sample was put in a 20 mL headspace bottle, incubated at 80 °C for 15 min, and then driven by N_2 into the chromatographic column for detection. The NIST (2020) and IMS databases (0.4.07) in the instrument software were used for a qualitative analysis of volatile compounds [33].

2.11. Data Analysis

The test results were repeated at least 3 times, and the final results were shown as mean ± standard deviation (SD). SPSS 24.0 (SPSS Inc., Chicago, IL, USA) was used for one-way ANOVA, the Duncan method was used for differences between groups, and $p < 0.05$ showed significant differences.

3. Results and Discussion

3.1. Effect of Fv-SDF Addition on Thermodynamic Properties of Wheat Starch

In order to better understand the effect of Fv-SDF on the quality of dried noodles, the thermodynamic properties of Fv-SDF on wheat starch were first studied. The thermodynamic properties can reflect the enthalpy change and moisture form of the mixed dough during crystallization and melting [34]. The peak temperature is the temperature at which the sample absorbs heat during gelatinization, the gelatinization enthalpy (ΔH) reflects the aggregation degree of protein and also shows the hydrophobic and hydrophilic properties of protein [35]. The results showed that the peak temperature, gelatinization temperature, and gelatinization enthalpy of the mix powder increased with the Fv-SDF addition ratio (Table 1 and Figure 1), but gelatinization enthalpy did not increase significantly. The initial gelatinization temperature T_0, peak temperature Tp, terminating gelatinization temperature Tc, and gelatinization enthalpy ΔH of the mix powder were 56.32 °C, 63.65 °C, 69.40 °C, and 1.80 W/g, respectively. With the increase of FV-SDF addition to 15%, the T_0, Tp, Tc, and ΔH of the mix powder increased to 61.55 °C, 67.12 °C, 73.26 °C, and 2.22 W/g, respectively. The increase in gelatinization temperature was consistent with the RVA results, indicating that a higher gelatinization temperature was required during the gelatinization process of wheat flour. This was due to the fact that Fv-SDF competed with proteins for water absorption, which led to the decrease of available water in proteins, thus increasing the energy needed for protein denaturation and increasing the gelatinization temperature of the mixture. These results were consistent with the findings of Wang et al. [36], which studied the effect of soluble soybean polysaccharides on starch gelatinization and found that the gelatinization temperature of the starch system increased, indicating that the addition of polysaccharides increased the difficulty of starch gelatinization.

Table 1. Effect of Fv-SDF addition on thermodynamic properties of wheat starch.

Fv-SDF Addition/%	Initial Gelatinization Temperature T_0/°C	Peak Temperature Tp/°C	Terminating Gelatinization Temperature Tc/°C	Gelatinization Enthalpy/(ΔH, W/g)
0	56.32 ± 0.23 ^{d}	63.65 ± 0.16 ^{e}	69.40 ± 0.35 ^{d}	1.80 ± 0.45 ^{b}
5.0	59.58 ± 0.16 ^{c}	65.38 ± 0.23 ^{d}	71.60 ± 0.95 ^{c}	2.14 ± 0.43 ^{a}
7.5	59.89 ± 0.08 ^{c}	65.89 ± 0.20 ^{c}	72.02 ± 1.10 ^{c}	2.29 ± 0.21 ^{a}
10.0	60.81 ± 0.33 ^{b}	66.56 ± 0.47 ^{b}	73.47 ± 1.85 ^{a}	2.30 ± 0.29 ^{a}
12.5	61.06 ± 0.42 ^{b}	66.66 ± 0.16 ^{b}	72.17 ± 0.33 ^{bc}	2.29 ± 0.14 ^{a}
15.0	61.55 ± 0.79 ^{a}	67.12 ± 0.23 ^{a}	73.26 ± 0.72 ^{ab}	2.22 ± 0.16 ^{a}

Note: Different letters represent significance in the same column ($p < 0.05$).

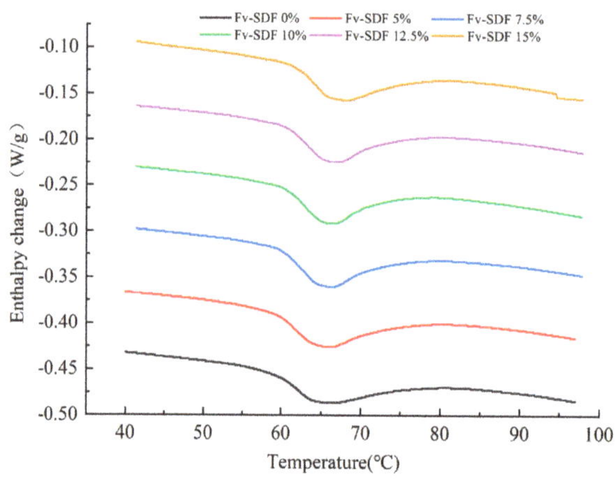

Figure 1. Effect of Fv-SDF addition on the thermodynamic properties of wheat starch.

3.2. Effect of Fv-SDF Addition on Pasting Properties of Wheat Starch

Secondly, the influence of Fv-SDF on the pasting properties of wheat starch was detected. The pasting properties mainly reflect the swelling ability of starch in the composite flour system and the binding ability of starch with water. In addition, it can also reflect the effect of starch gelatinization characteristics on flour products [37]. As shown in Figure 2, the viscosity of starch paste gradually increased with the extension of heating time; the viscosity peak appeared at 300–400 s and moved down with the increase of Fv-SDF addition. With the further extension of heating time, the viscosity of the starch paste decreased and the viscosity trough appeared at 450–600 s. With the increase of Fv-SDF addition, the viscosity trough moved down. When the heating time was 450–600 s, the fluidity of the starch paste decreased, while the viscosity increased.

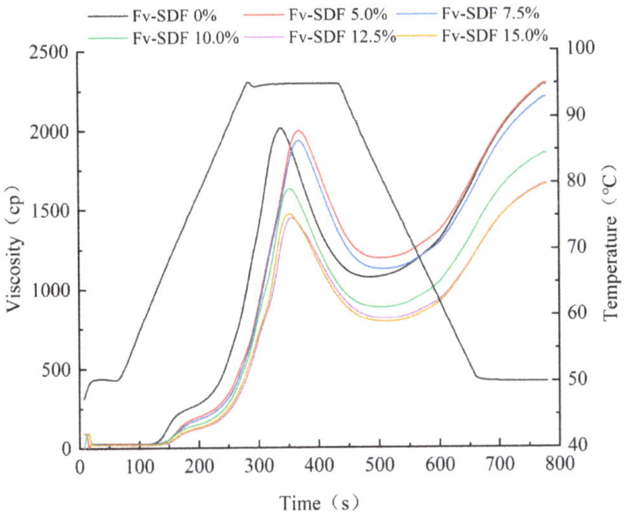

Figure 2. Effect of Fv-SDF addition on the pasting properties of wheat starch.

The essence of starch gelatinization is that water enters the crystallization region of starch particles and the internal structure of starch changes from ordered to disordered [38]. As shown in Table 2, with the increase of Fv-SDF content, the viscosity and setback value of the mix powder showed a downward trend. The decrease in viscosity may be due to the fact that the addition of Fv-SDF diluted the starch content of wheat and reduced the cross-linking degree of the starch structure in the mixed paste, resulting in a decrease in viscosity of the mixture [39]. In addition, the amylose content also contributed to the reduction of viscosity, which was consistent with other research results [40]. The setback value is an important indicator of short-term retrogradation/aging of amylose [41]. Fv-SDFs, starches, and proteins interacted to form new macromolecules, which hindered the interaction between starch molecules and delayed the formation of crystals. The breakdown value reflects the stability of the starch paste, and Fv-SDF improved the thermal stability of the mixture. Our results were consistent with the findings of Wang et al. [42], that is, adding polysaccharide to flour can reduce viscosity.

Table 2. Effect of Fv-SDF addition on the pasting properties of wheat starch.

Fv-SDF Addition/%	Peak Viscosity/cp	Trough Viscosity/cp	Breakdown Value/cp	Final Viscosity/cp	Setback Value/cp	Peak Time/min	Pasting Temperature/°C
0	2015.0 ± 14.24^{a}	1069.5 ± 23.54^{a}	945.5 ± 7.78^{a}	2250.0 ± 50.91^{a}	1180.5 ± 47.38^{a}	5.8 ± 0.14^{a}	64.8 ± 1.87^{e}
5.0	1962.5 ± 47.38^{ab}	1168.5 ± 30.41^{a}	809.0 ± 4.24^{b}	2237.0 ± 74.95^{a}	1106.0 ± 8.49^{b}	6.1 ± 0.86^{a}	69.3 ± 1.00^{d}
7.5	1876.0 ± 82.02^{b}	1031.5 ± 30.81^{a}	844.5 ± 48.79^{b}	1972.0 ± 38.10^{ab}	940.5 ± 87.28^{c}	5.9 ± 0.37^{a}	76.9 ± 0.47^{c}

Table 2. Cont.

Fv-SDF Addition/%	Peak Viscosity/cp	Trough Viscosity/cp	Breakdown Value/cp	Final Viscosity/cp	Setback Value/cp	Peak Time/min	Pasting Temperature/°C
10.0	1595.0 ± 46.67 c	845.0 ± 52.33 b	750.0 ± 5.66 c	1706.0 ± 25.06 b	861.0 ± 52.74 cd	5.8 ± 0.10 a	86.2 ± 1.34 b
12.5	1481.0 ± 24.04 c	850.0 ± 53.74 b	631.0 ± 5.66 d	1689.5 ± 42.43 b	839.0 ± 11.31 d	6.0 ± 0.10 a	88.0 ± 1.23 a
15.0	1488.0 ± 48.08 c	799.5 ± 12.02 b	688.5 ± 12.02 d	1669.5 ± 27.58 b	870.0 ± 15.56 cd	5.9 ± 0.04 a	87.6 ± 0.53 ab

Note: Different letters represent significance in the same column ($p < 0.05$).

3.3. Effect of Fv-SDF Addition on Farinograph Properties of Mix Powder

The farinograph properties of wheat flour reflect the changes in rheological characteristics during dough mixing and are one of the important indexes to evaluate the processing characteristics of wheat flour [43]. The farinograph properties of the mixture of Fv-SDF and wheat flour are shown in Table 3. With the increase of Fv-SDF addition, the dough stability time, dough development time, and flour quality index increased, while the water absorption and weakening degree decreased. The stability time of dough is closely related to the toughness and gluten strength of the dough [44]. Our results showed that the addition of Fv-SDF could prolong the stability time of dough, improve the mechanical stirring resistance of dough, and enhance gluten strength. In addition, Fv-SDF is a kind of gel polysaccharide, which can connect the gluten network structures, delay the breaking of disulfide bonds in the dough, and the depolymerization of gluten macromolecules, making the gluten network more stable [45]. The dough development time refers to the time required for stirring from the beginning of adding water to the maximum consistency of dough [46]. With the increase of the dosage of Fv-SDF, the dough development time tended to increase, which was because the high water-holding capacity of Fv-SDF reduced the water absorption of the gluten protein, prolonged the dough formation time, and further affected the formation rate of the gluten network. These results were similar to those obtained by adding bean dregs, apple dregs, and oats to wheat flour [47]. The decrease in water absorption of dough caused by Fv-SDF was mainly due to the fact that Fv-SDF was more hydrophilic than starch. Fv-SDF could form a barrier layer around the starch granules, which prevented water molecules from contacting the starch granules and limited the expansion of the starch granules. In addition, previous studies also reported that adding a high content of sugar and fiber to wheat flour could increase the stability time of dough and decrease water absorption [48]. To sum up, adding Fv-SDF to wheat flour was beneficial to the formation of a gluten network, which could not only improve gluten strength but also enhance the nutritional characteristics of the mixed flour. Therefore, Fv-SDF could be used as an improver for medium and low gluten flours.

Table 3. Effect of Fv-SDF addition on the farinogram properties of mix powder.

Fv-SDF Addition/%	Consistency/ FU	Water Absorption (Adjust 500 FU/%)	Dough Development Time/min	Stability Time/min	Weakening Degree (Firs 10 min/FU)	Weakening Degree (ICC/after Reaching the Maximum 12 min)	Flour Quality Index/mm
0	491.3 ± 3.18 a	62.2 ± 1.52 a	3.2 ± 0.14 c	2.8 ± 0.39 d	93.5 ± 6.36 ab	126.5 ± 7.78 a	48.0 ± 1.41 e
5.0	505.0 ± 5.66 a	55.1 ± 1.38 b	3.5 ± 0.25 bc	3.3 ± 0.25 d	109.3 ± 16.62 a	137.5 ± 14.85 a	52.3 ± 1.77 de
7.5	507.8 ± 15.20 a	51.8 ± 0.74 c	3.7 ± 0.04 bc	3.8 ± 0.42 cd	92.0 ± 12.73 ab	120.3 ± 16.62 ab	57.0 ± 1.41 d
10.0	497.5 ± 0.71 a	49.2 ± 0.07 d	4.5 ± 0.39 ab	4.7 ± 0.53 bc	78.8 ± 6.01 bc	111.8 ± 7.42 abc	66.5 ± 0.71 c
12.5	490.0 ± 12.73 a	46.6 ± 0.12 e	4.9 ± 0.21 a	5.5 ± 0.71 ab	59.5 ± 7.78 cd	97.3 ± 10.96 bc	75.75 ± 5.30 b
15.0	500.0 ± 7.07 a	44.5 ± 0.85 e	5.6 ± 0.88 a	6.7 ± 0.57 a	43.5 ± 4.95 d	88.0 ± 2.83 c	88.0 ± 4.24 a

Note: Different letters represent significance in the same column ($p < 0.05$).

3.4. Effect of Fv-SDF Addition on Dynamic Rheological Properties of Dough

The dynamic rheological properties of dough mainly refer to its viscoelasticity. The viscoelasticity of dough could characterize its structure and physicochemical properties, that is, the tight combination of moisture and gluten, thus predicting the changes in dough processing and the quality of products [49]. As shown in Figure 3, with the increase of angular frequency, the storage modulus (G′) (Figure 3A) and the loss modulus (G″) (Figure 3B) of the samples showed an upward trend. The increase in dynamic modulus was due to the fact that the hydroxyl groups in Fv-SDF changed the moisture distribution of the dough. Besides, Fv-SDF crosslinked with the protein in wheat flour to form a relatively stable gel structure, which enhanced the elasticity of gluten protein and led to the increase of G′ and G″ [50]. In addition, the G′ of the same sample was obviously higher than that of G″, and the overall performance was elasticity [51]. When the addition of Fv-SDF was 10%, the viscoelasticity of the dough was better and the quality of the dried noodles was excellent. As shown in Figure 3C, with the increase of Fv-SDF addition, the values of loss tangent (tanδ = G″/G′) showed an upward trend. However, the tanδ values were all less than 1, indicating that the elasticity of the dough was dominant. Those results indicated that the Fv-SDF mixed dough system was more stable and the micro-fermented dried noodles were not easy to break in the low-frequency scanning range. To sum up, the addition of Fv-SDF can increase the ductility of the dough.

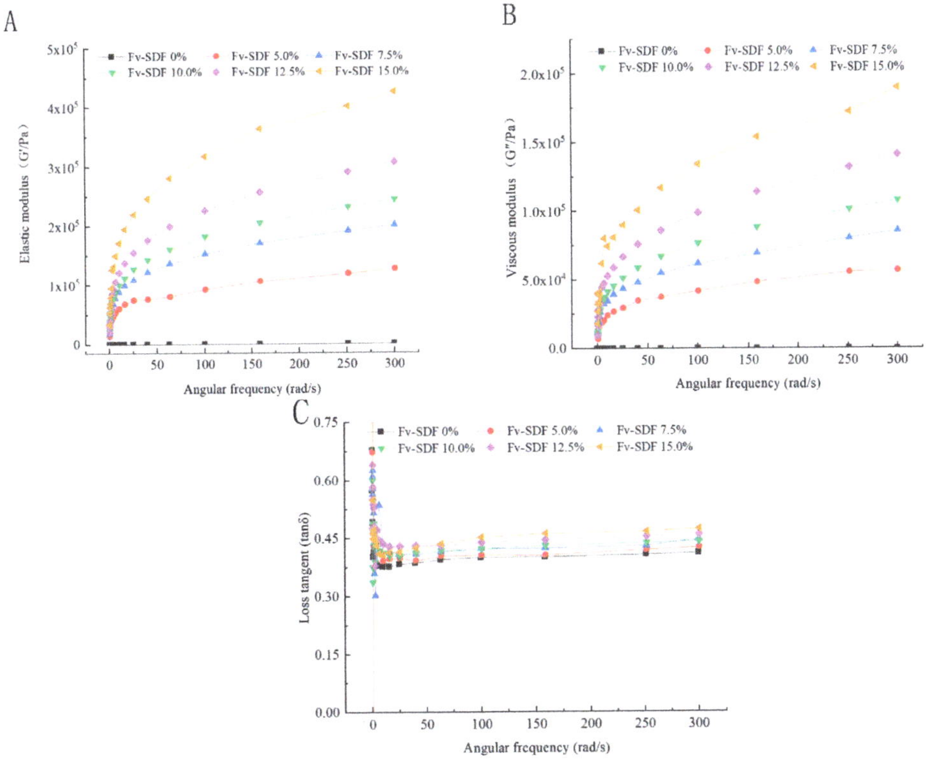

Figure 3. Effects of Fv-SDF addition on the storage modulus (G′) (**A**), loss modulus (G″) (**B**), and loss tangent (tanδ = G″/G′) (**C**) of dough.

3.5. Effect of Fv-SDF Addition on Apparent Color of Dried Noodles

The effect of Fv-SDF addition on the whiteness of micro-fermented dried noodles was determined by colorimeter. Where L* is the whiteness, with a value range of 0–100, and the greater the value, the whiter the color. a* is the red-green value, the negative value represents green, and the positive value represents red. b* is the yellow-blue value, the negative value indicates that the color is blue, and the positive value indicates that the color is yellow. As shown in Table 4, compared with the control group, with the increase of Fv-SDF addition, the whiteness value L* of Fv-SDF micro-fermented dried noodles decreased significantly, while a* and b* increased significantly. The L* value of Fv-SDF powder was the lowest, and the values of a* and b* were the highest. Among them, L*, a*, and b* were all positive values, indicating that the whiteness of Fv-SDF micro-fermented dried noodles was reduced and that it was reddish and yellowish. Compared with the high Fv-SDF addition, the effect of low Fv-SDF addition on the apparent color of micro-fermented dried noodles was more obvious. In summary, when the addition of FV-SDF was 10%, the appearance of micro-fermented dried noodles was more pleasant.

Table 4. Effect of Fv-SDF addition on the apparent color of dried noodles.

Fv-SDF Addition/%	L*	a*	b*
100.0 (Fv-SDFpowder)	38.21 ± 7.75 [f]	12.75 ± 0.40 [a]	27.88 ± 3.92 [a]
0	88.25 ± 1.47 [a]	0.19 ± 0.05 [e]	9.89 ± 0.34 [d]
5.0	64.32 ± 0.86 [b]	5.51 ± 0.41 [d]	13.61 ± 0.78 [c]
7.5	58.34 ± 1.51 [c]	6.56 ± 0.78 [c]	13.74 ± 0.39 [c]
10.0	52.25 ± 0.65 [d]	8.92 ± 0.56 [b]	17.18 ± 0.98 [b]
12.5	47.80 ± 1.04 [e]	9.53 ± 0.38 [b]	17.23 ± 0.48 [b]
15.0	46.84 ± 0.67 [e]	9.56 ± 0.22 [b]	17.52 ± 0.79 [b]

Note: Different letters represent significance in the same column ($p < 0.05$).

3.6. *Effect of Fv-SDF Addition on the Cooking Quality of Dried Noodles*

The best optimum cooking time is the time when the white core of the noodles disappears completely during steaming [52]. As shown in Figure 4A, with the increase of the Fv-SDF addition, the optimum cooking time of Fv-SDF micro-fermented dried noodles gradually decreased. The addition of Fv-SDF diluted the starch content, and the decrease in starch content led to a decrease in gelatinization temperature. After micro-fermentation, there were many air holes in the noodles, and the water molecules could easily enter the inside, which made the starch absorb water and swell rapidly, reducing the optimal cooking time [53]. When the addition of Fv-SDF was 10%, the optimum cooking time of dried noodles was 142 s. The addition of Fv-SDF had a significant effect on the water absorption and cooking loss rate of micro-fermented dried noodles. Our results showed that when the addition of Fv-SDF was 10%, the dry matter water absorption of micro-fermented dried noodles was the highest and the cooking loss rate of micro-fermented dried noodles remained at a low level (Figure 4B). With the further increase of Fv-SDF addition, the water absorption of micro-fermented dried noodles showed a downward trend, while the cooking loss rate showed an upward trend. FV-SDF had high water absorption, which was able to absorb water and form a gelatinous substance, and lock the moisture in the noodles, making the noodles absorb water and mature faster. The addition of Fv-SDF could form a protective net around the starch particles, thus reducing the spontaneous rupture of expanded particles, reducing the dissolution rate of starch, and reducing the loss rate of micro-fermented dried noodles during the cooking.

The results of the sensory evaluation are shown in Figure 5. With the increase of the addition of Fv-SDF, the sensory score of Fv-SDF micro-fermented dried noodles generally showed a trend of increasing first and then reducing. When the addition of Fv-SDF was less than 10%, the toughness and taste scores were higher. The addition of Fv-SDF enhanced the gluten strength, chewiness, and elasticity of Fv-SDF micro-fermented dried noodles, and at the same time made the noodles have mushroom flavor and micro-fermentation aroma. In addition, the appearance of Fv-SDF micro-fermented dried noodles was uniform, and the color gradually darkened, giving it an attractive caramel color. The sensory score of micro-fermented dried noodles decreased with the further increase of Fv-SDF addition, because excessive addition of Fv-SDF reduced the hardness, chewiness, and toughness, and the color of micro-fermented dried noodles was dark brown, resulting in a decrease in the sensory evaluation of Fv-SDF micro-fermented dried noodles. When the addition of Fv-SDF was 10%, the chewiness, elasticity, flavor, and color of the micro-fermented dry noodles were better, and the sensory scores of the micro-fermented dry noodles were the highest.

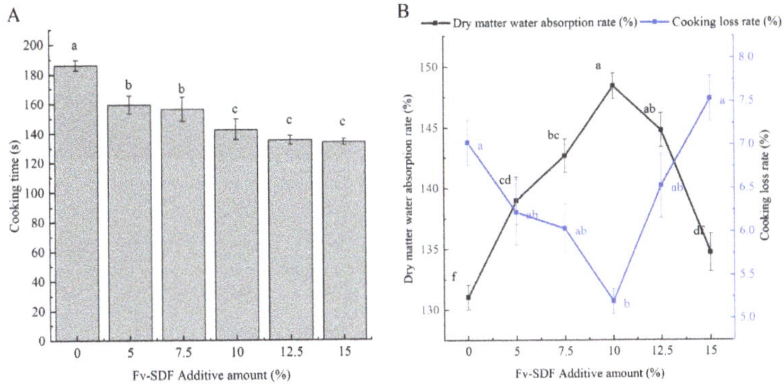

Figure 4. The effect of Fv-SDF addition on the optimum cooking time (**A**), water absorption, and cooking loss rate (**B**) of dried noodles. Different letters represent significance ($p < 0.05$).

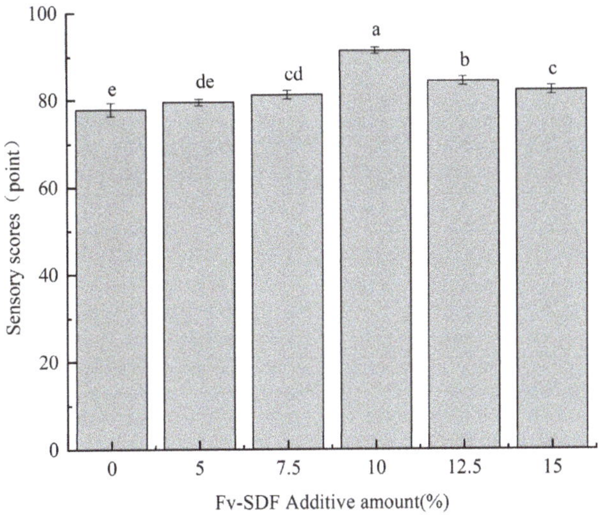

Figure 5. Effects of Fv-SDF addition on sensory evaluation of dried noodles. Different letters represent significance ($p < 0.05$).

3.7. Comparative Analysis of Volatile Flavor Compounds in Dried Noodles

In order to further clarify the differences in volatile flavor compounds in dried noodles, the GC-IMS was selected to identify the characteristic peak areas of the samples. The three dried noodle samples were as follows: CK group (unfermented, Fv-SDF addition amount 0%), Fv-SDF 0% (micro-fermentation for 9 min, Fv-SDF addition amount 0%), and Fv-SDF 10% (micro-fermentation for 9 min, Fv-SDF addition amount 10%).

In the three-dimensional spectrum (Figure 6A) and two-dimensional spectral top view (Figure 6B,C) of volatile flavor compounds of dried noodles, obvious visual differences of volatile flavor compounds could be observed among different samples (CK, 0% and 10%) in the GC-IMS spectrum.

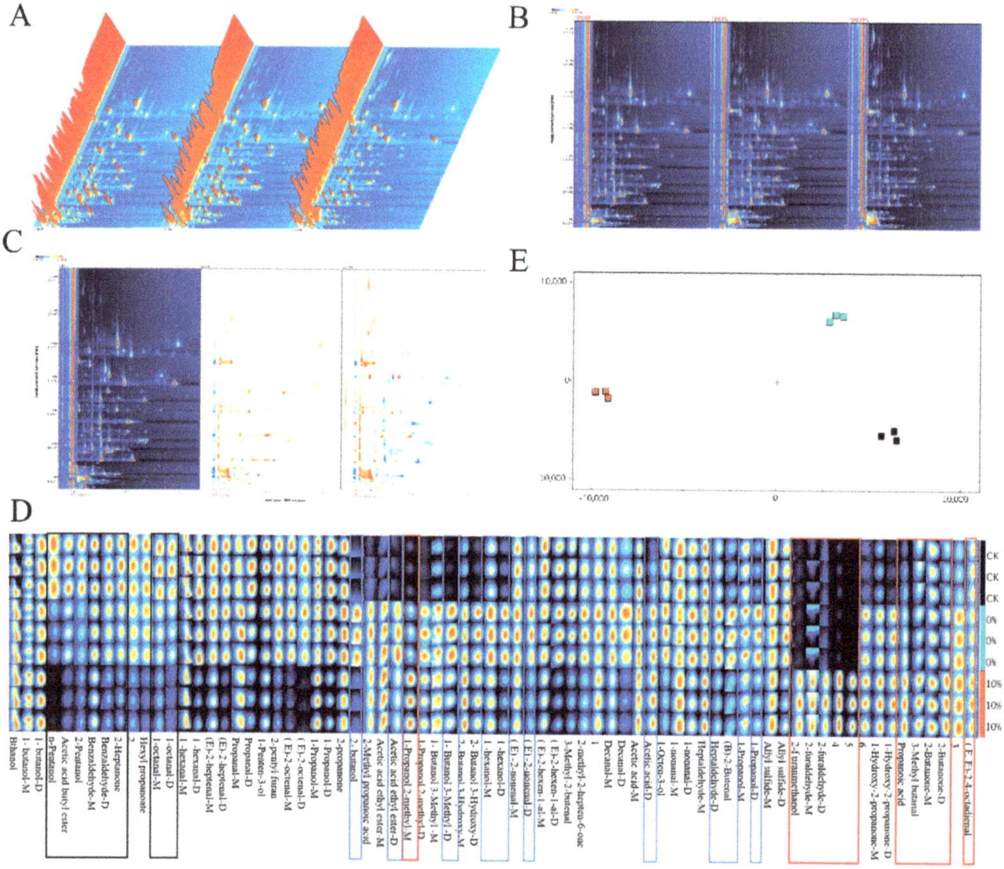

Figure 6. Three-dimensional spectrum of volatile substance composition (**A**), 2D spectrum of volatile substance composition (**B**), 2D differential spectrum of volatile substance composition (**C**), GC-IMS fingerprint (**D**), PCA analysis (**E**). In (**A–C**), the samples from left to right are the CK group, Fv-SDF 0% group, and Fv-SDF 10% group, respectively.

Figure 6D was a fingerprint of the volatile components in three samples of dried noodles, which represented the signal peak of each volatile substance in different samples. The darker the red color, the higher the concentration of the substance. By comparing and analyzing the volatile components in three samples of dried noodles, the results showed that there was no significant difference in the composition and content of flavor components between the CK group and the Fv-SDF 0% group, but there was a significant difference in the composition and content of flavor components between the CK group and Fv-SDF 10% group. It could be seen from Figure 6D that there was little difference in the content of ethanol and butanol in dried noodles. It could be seen from the red frame that compared with the CK group and Fv-SDF 0% group, the contents of furfural, 1-hydroxy-2-acetone, propionic acid, and 3-methylbutyraldehyde increased in the Fv-SDF 10% group. Among them, furfural, 1-hydroxy-2-acetone, and 3-methylbutyraldehyde provided caramel flavor and sweetness, and propionic acid provided yogurt and vinegar flavors. In addition, propionic acid may be the sour taste produced by the micro-fermentation, while furfural, 1-hydroxy -2-acetone, and 3-methylbutyraldehyde were produced by non-starch polysaccharides contained in Fv-SDF. In the blue frame, it could be seen that compared with the CK group and Fv-SDF 10% group, the contents of 2-butanol, 2-methylpropionic acid, ethyl acetate,

2-methylpropanol, 3-methylbutanol, 3-hydroxy-2-butanonend were higher in Fv-SDF 0% group. Among them, 2-methylpropionic acid provided a sour taste, ethyl acetate provided a fresh taste, and 2-methylpropionic acid provided a sour taste which was produced during the micro-fermentation. As can be seen from the black frame in Figure 6D, compared with the Fv-SDF 0% group and Fv-SDF 10% group, the contents of pentanal, butyl acetate, benzaldehyde, 2-heptanone, hexyl propionate, octanal, hexanal, and trans-2-heptenal were higher, which provided a clear fragrance, fruity flavor, and fatty taste. Compared with the CK group, the content of flavor substances in the Fv-SDF 0% group was higher, and the types were richer, which were mainly produced by micro-fermentation. Among them, 3-methylbutyraldehyde has a strong malt and yeast flavor, which is a product of protein hydrolysis and amino acid (isoleucine and leucine) degradation, and it is an important flavor compound in many fermentation products [54].

The PCA analysis of three samples was shown in Figure 6E, which is a dimensionality reduction method commonly used to reduce the dimension of large data sets. The results showed that the contribution rates of PC-1 and PC-2 were 63% and 34%, respectively, and the cumulative variance contribution rate was 97%, which indicated that the PCA results were reliable. The distribution of the PC-2 axis of the CK, Fv-SDF 0%, and Fv-SDF 10% group was significantly different. The CK and Fv-SDF 0% groups were close to each other on the PC-1 axis, and these two groups were farther away from the Fv-SDF 10% group on the PC-1 axis. There was a great difference in flavor between the CK and Fv-SDF 10% group, while there was a little difference between the Fv-SDF 0% and 10% group, which was consistent with the fingerprint analysis.

In order to further clarify the composition of flavor compounds in the three dried noodle samples, all flavor compounds were classified. As shown in Table 5, 67 volatile substances, including 20 alcohols, 21 aldehydes, 9 ketones, 4 esters, and 13 other compounds, were detected in all three samples. There were significant differences in the volatile components of the three dried noodle samples, but the flavor components in the CK and Fv-SDF 10% groups were more significant. Consistent with fingerprint analysis, the contents of furfural, 1-hydroxy-2-acetone, propionic acid, and 3-methylbutyraldehyde were higher in the Fv-SDF 10% group.

Table 5. Analysis of volatile flavor compounds in three samples of dried noodles.

	Name	CAS	Rt [s]	Molecular	Odor Description
Alcohols	1-octene-3-ol	C3391864	867.68	$C_8H_{16}O$	Mushroom aroma
	1-hexanol-M	C111273	713.00	$C_6H_{14}O$	Fresh, fruity, alcohol,
	1-hexanol-D	C111273	712.41	$C_6H_{14}O$	Fresh, fruity, alcohol,
	1-Pentanol-M	C71410	533.20	$C_5H_{12}O$	balm
	1-Pentanol-D	C71410	532.28	$C_5H_{12}O$	balm
	1-butanol, 3-methyl-M	C123513	457.43	$C_5H_{12}O$	Whisky, banana fruit
	1-butanol, 3-methyl-D	C123513	456.99	$C_5H_{12}O$	Whisky, banana fruit
	1-pentene-3-ol	C616251	388.08	$C_5H_{10}O$	Ethereal, green, tropical fruit
	1-butanol-M	C71363	369.98	$C_4H_{10}O$	Red wine taste
	1-butanol-D	C71363	369.04	$C_4H_{10}O$	Red wine taste
	1-propanol, 2-methyl-M	C78831	311.92	$C_4H_{10}O$	Fresh, wine, leather
	1-propanol, 2-methyl-D	C78831	310.98	$C_4H_{10}O$	Fresh, wine, leather
	1-propanol-M	C71238	260.59	C_3H_8O	Alcohol, pungent odor
	1-propanol-D	C71238	259.72	C_3H_8O	Alcohol, pungent odor
	Ethyl alcohol	C64175	191.40	C_2H_6O	toasty
	2-fluoro-uranyl alcohol	C98000	1358.68	$C_5H_6O_2$	toasty
	1-octanol-M	C124130	609.80	$C_8H_{16}O$	Aldehyde, waxy, fruity, fat
	1-octanol-D	C124130	609.23	$C_8H_{16}O$	Aldehyde, waxy, fruity, fat
	2-Pentanol	C6032297	339.50	$C_5H_{12}O$	Fusel oil, fragrant
	2-butanol	C78922	228.99	$C_4H_{10}O$	fruity

Table 5. *Cont.*

	Name	CAS	Rt [s]	Molecular	Odor Description
Aldehyde	Benzaldehyde -M	C100527	957.31	C_7H_6O	Bitter almond, cherry and nut notes
	Benzaldehyde D	C100527	957.31	C_7H_6O	Bitter almond, cherry and nut notes
	(E)-2-octenal -M	C2548870	809.46	$C_8H_{14}O$	Fresh cucumber, fat, green fragrance
	(E)-2-octenal-D 1-nonaldehyde-D	C2548870	807.91	$C_8H_{14}O$	Fresh cucumber, fat, green fragrance
	1-nonanal-M	C124196	759.05	$C_9H_{18}O$	Rose, citrus, etc
	1-nonaldehyde-D	C124196	759.05	$C_9H_{18}O$	Rose, citrus, etc
	(E)-2-heptenaldehyde-M	C18829555	660.72	$C_7H_{12}O$	Green fragrant vegetables, fresh, fat
	(E)-2-heptenaldehyd-D	C18829555	662.94	$C_7H_{12}O$	Green fragrant vegetables, fresh, fat
	(E)-2-hexene-1-aldehyde-M	C6728263	476.27	$C_6H_{10}O$	Green fragrant vegetables, fresh, fat
	(E)-2-hexene-1-aldehyde-D	C6728263	476.27	$C_6H_{10}O$	Green fragrant vegetables, fresh, fat
	3-methyl-2-butenal	C107868	447.79	C_5H_8O	fruity
	Heptaldehyde -M	C111717	424.13	$C_7H_{14}O$	Fresh, fat, green and fruity
	Heptyl aldehyde D	C111717	425.00	$C_7H_{14}O$	Fresh, aldehyde, fat, green,
	1-hexaldehyde-M	C66251	302.95	$C_6H_{12}O$	Fresh, green, fat,
	1-hexaldehyde-D	C66251	302.16	$C_6H_{12}O$	Fresh, green, fat,
	(E)-2-butenal	C123739	267.24	C_4H_6O	null
	n-valeraldehyde	C110623	221.23	$C_5H_{10}O$	Grass smell, taste exciting
	Propanal-M	C123386	134.93	C_3H_6O	Pungent smell, grass smell
	Propional-D	C123386	135.47	C_3H_6O	Pungent smell, grass smell
	3-methylbutyraldehyde	C590863	182.31	$C_5H_{10}O$	Chocolate, fat
	(E, E)-2, 4-octanedienal	C30361285	1256.70	$C_8H_{12}O$	Fat, green, pear, melon
	2-furfural-M	C98011	878.87	$C_5H_4O_2$	Sweet, wood, toast
	2-furfural-D	C98011	879.98	$C_5H_4O_2$	Sweetness, wood, bread
Ketone	2-methyl-2-heptene-6-one	C110930	683.56	$C_8H_{14}O$	Citrus, fruity, keto
	1-hydroxy-2-acetone-M	C116096	628.11	$C_3H_6O_2$	Pungent smell, caramel taste
	1-hydroxy-2-acetone-D	C116096	628.57	$C_3H_6O_2$	Pungent smell, caramel taste
	3-hydroxy-2-butanone-M	C513860	599.09	$C_4H_8O_2$	Buttery, creamy
	2-butanone, 3-hydroxy-D	C513860	599.09	$C_4H_8O_2$	Buttery, creamy
	2-heptanone	C110430	411.30	$C_7H_{14}O$	Pear, banana like fruit
	2-butanone-M	C78933	176.41	C_4H_8O	Fruity, camphor aromas
	2-butanone-D	C78933	177.48	C_4H_8O	Fruity, camphor aromas
	2-acetone	C67641	144.05	C_3H_6O	Fresh, apple, pear
Esters	Hexyl propionate	C2445763	671.85	$C_9H_{18}O_2$	Sweet fruity smell, soil fragrance
	Butyl acetate	C123864	287.17	$C_6H_{12}O_2$	fruity
	Ethyl acetate -M	C141786	169.26	$C_4H_8O_2$	Fresh, fruity, sweet
	Ethyl acetate -D	C141786	169.08	$C_4H_8O_2$	Fresh, fruity, sweet

4. Conclusions

In this study, the effects of Fv-SDF on dough processing characteristics and micro-fermented dried noodles were investigated. The farinograph and rheological properties tests showed that the addition of Fv-SDF was positively correlated with dough stability time, development time, and the flour quality index, but negatively correlated with dough water absorption and weakening degree. When the addition of FV-SDF was 15%, the dough stability time was 6.7 min and the development time reached 5.6 min. Fv-SDF could improve the elasticity of dough and enhance the strength of gluten. RVA results showed that the retrogradation value of the mixed flour decreased with the increase of Fv-SDF content, which indicated that the addition of Fv-SDF had an anti-aging effect and prolonged the shelf life of the noodles. The addition of 10% Fv-SDF made noodles more acceptable with the highest sensory score of 92, including pleasant color and good cooking characteristics. When the addition of Fv-SDF was 10%, the optimal cooking time was 142 s, the water absorption rate of dried noodles reached 148%, and the loss rate reached 5.2%. The results of GC-IMS analysis showed that the unique source of flavor compounds of noodles was generated by Fv-SDF and micro-fermentation treatment. Adding 10% Fv-SDF not only improved the nutritional value of the noodles but also gave them a unique flavor. This research can provide technical support for the postharvest processing and product development of *Flammulina velutipes* and provide a technical reference and theoretical basis for the application of Fv-SDF in micro-fermented dried noodle products.

Supplementary Materials: The following supporting information can be downloaded at: https://www.mdpi.com/article/10.3390/foods13172764/s1, Table S1. Sensory evaluation.

Author Contributions: Conceptualization, writing—original draft, writing—review and editing, W.W. and S.Y.; conceptualization, writing—original draft, writing—review and editing, F.J.; methodology, data curation, resources, L.W.; methodology, data curation, F.H. and S.S.; supervision, project administration, Y.W. and Z.G.; project administration, funding acquisition, F.J. All authors have read and agreed to the published version of the manuscript.

Funding: This project was funded by the Natural Science Foundation of Shandong Province (ZR2022MC096), the Natural Science Foundation of Shandong Province (ZR2022QC239), and the Shandong Province Modern Agricultural Industry Technology System (SDAIT-07-09).

Institutional Review Board Statement: Not applicable.

Informed Consent Statement: Not applicable.

Data Availability Statement: The original contributions presented in the study are included in the article/Supplementary Material, further inquiries can be directed to the corresponding author.

Acknowledgments: The authors would like to acknowledge the support of the Natural Science Foundation of Shandong Province and Shandong Province Modern Agricultural Industry Technology System.

Conflicts of Interest: The authors declare no conflicts of interest.

References

1. Yeh, M.Y.; Ko, W.C.; Lin, L.Y. Hypolipidemic and antioxidant activity of enoki mushrooms (*Flammulina velutipes*). *Biomed. Res. Int.* **2014**, *2014*, 352385. [CrossRef] [PubMed]
2. Liu, X.B.; Xia, E.H.; Li, M.; Cui, Y.Y.; Wang, P.M.; Zhang, J.X.; Xie, B.G.; Xu, J.P.; Yan, J.J.; Li, J.; et al. Transcriptome data reveal conserved patterns of fruiting body development and response to heat stress in the mushroom-forming fungus *Flammulina filiformis*. *PLoS ONE* **2020**, *15*, e0239890. [CrossRef]
3. Wei, Q.; Pan, X.; Li, J.; Jia, Z.; Fang, T.; Jiang, Y. Isolation and Molecular Identification of the Native Microflora on *Flammulina velutipes* Fruiting Bodies and Modeling the Growth of Dominant Microbiota (*Lactococcus lactis*). *Front. Microbiol.* **2021**, *12*, 664874. [CrossRef]
4. Wang, R.; Zhang, Y.; Lu, H.; Liu, J.; Song, C.; Xu, Z.; Yang, H.; Shang, X.; Feng, T. Comparative Aroma Profile Analysis and Development of a Sensory Aroma Lexicon of Seven Different Varieties of *Flammulina velutipes*. *Front. Nutr.* **2022**, *9*, 827825. [CrossRef]
5. Tsai, S.-Y.; Hwang, B.-F.; Wang, Y.-H.; Lin, C.-P. Moisture desorption and thermal properties of polysaccharide from pulsed light irradiated *Flammulina velutipes*. *J. Therm. Anal. Calorim.* **2017**, *127*, 469–481. [CrossRef]
6. Liang, Q.; Zhao, Q.; Hao, X.; Wang, J.; Ma, C.; Xi, X.; Kang, W. The Effect of *Flammulina velutipes* Polysaccharide on Immunization Analyzed by Intestinal Flora and Proteomics. *Front. Nutr.* **2022**, *9*, 841230. [CrossRef] [PubMed]
7. Banerjee, D.K.; Das, A.K.; Banerjee, R.; Pateiro, M.; Nanda, P.K.; Gadekar, Y.P.; Biswas, S.; McClements, D.J.; Lorenzo, J.M. Application of Enoki Mushroom (*Flammulina velutipes*) Stem Wastes as Functional Ingredients in Goat Meat Nuggets. *Foods* **2020**, *9*, 432. [CrossRef]
8. Cao, Y.; Tian, B.; Zhang, Z.; Yang, K.; Cai, M.; Hu, W.; Guo, Y.; Xia, Q.; Wu, W. Positive effects of dietary fiber from sweet potato [*Ipomoea batatas* (L.) Lam.] peels by different extraction methods on human fecal microbiota in vitro fermentation. *Front. Nutr.* **2022**, *9*, 986667. [CrossRef]
9. Bai, X.; He, Y.; Quan, B.; Xia, T.; Zhang, X.; Wang, Y.; Zheng, Y.; Wang, M. Physicochemical properties, structure, and ameliorative effects of insoluble dietary fiber from tea on slow transit constipation. *Food Chem. X* **2022**, *14*, 100340. [CrossRef]
10. Li, S.; Hu, N.; Zhu, J.; Zheng, M.; Liu, H.; Liu, J. Influence of modification methods on physicochemical and structural properties of soluble dietary fiber from corn bran. *Food Chem. X* **2022**, *14*, 100298. [CrossRef]
11. Thomas, M.S.; Calle, M.; Fernandez, M.L. Healthy plant-based diets improve dyslipidemias, insulin resistance, and inflammation in metabolic syndrome. A narrative review. *Adv. Nutr.* **2023**, *14*, 44–54. [CrossRef] [PubMed]
12. Coţovanu, I.; Mironeasa, C.; Mironeasa, S. Nutritionally Improved Wheat Bread Supplemented with Quinoa Flour of Large, Medium and Small Particle Sizes at Typical Doses. *Plants* **2023**, *12*, 1698. [CrossRef]
13. Weng, J.; Chen, M.; Shi, B.; Liu, D.; Weng, S.; Guo, R. Konjac glucomannan defends against high-fat diet-induced atherosclerosis in rabbits by promoting the PI3K/Akt pathway. *Heliyon* **2023**, *9*, e13682. [CrossRef] [PubMed]

14. Qi, J.; Gao, J.; Zhang, Y.; Hou, W.; Han, T.; Sun, C. The Association of Dietary Fiber Intake in Three Meals with All-Cause and Disease-Specific Mortality among Adults: The U.S. National Health and Nutrition Examination Survey, 2003–2014. *Nutrients* **2022**, *14*, 2521. [CrossRef] [PubMed]
15. Elleuch, M.; Bedigian, D.; Roiseux, O.; Besbes, S.; Blecker, C.; Attia, H. Dietary fibre and fibre-rich by-products of food processing: Characterisation, technological functionality and commercial applications: A review. *Food Chem.* **2011**, *124*, 411–421. [CrossRef]
16. Pankiewicz, U.; Zielińska, E.; Sobota, A.; Wirkijowska, A. The Use of Saccharomyces cerevisiae Supplemented with Intracellular Magnesium Ions by Means of Pulsed Electric Field (PEF) in the Process of Bread Production. *Foods* **2022**, *11*, 3496. [CrossRef]
17. Huber, E.; Francio, D.L.; Biasi, V.; Mezzomo, N.; Ferreira, S.R. Characterization of vegetable fiber and its use in chicken burger formulation. *J. Food Sci. Technol.* **2016**, *53*, 3043–3052. [CrossRef]
18. Iversen, K.N.; Jonsson, K.; Landberg, R. The Effect of Rye-Based Foods on Postprandial Plasma Insulin Concentration: The Rye Factor. *Front. Nutr.* **2022**, *9*, 868938. [CrossRef]
19. Lin, Q.; Ren, A.; Liu, R.; Xing, Y.; Yu, X.; Jiang, H. Flavor properties of Chinese noodles processed by dielectric drying. *Front. Nutr.* **2022**, *9*, 1007997. [CrossRef]
20. Adamczyk, G.; Posadzka, Z.; Witczak, T.; Witczak, M. Comparison of the Rheological Behavior of Fortified Rye-Wheat Dough with Buckwheat, Beetroot and Flax Fiber Powders and Their Effect on the Final Product. *Foods* **2023**, *12*, 559. [CrossRef]
21. Hu, Z.; Guo, W.; Liu, C.; Wang, X.; Hong, J.; Liu, M.; Sun, B.; Zheng, X. Effect of polysaccharide on rheology of dough, microstructure, physicochemical properties and quality of fermented hollow dried noodles. *LWT* **2024**, *200*, 116214. [CrossRef]
22. Yan, J.K.; Wu, L.X.; Cai, W.D.; Xiao, G.S.; Duan, Y.; Zhang, H. Subcritical water extraction-based methods affect the physicochemical and functional properties of soluble dietary fibers from wheat bran. *Food Chem.* **2019**, *298*, 124987. [CrossRef] [PubMed]
23. Ginindza, A.; Solomon, W.K.; Shelembe, J.S.; Nkambule, T.P. Valorisation of brewer's spent grain flour (BSGF) through wheat-maize-BSGF composite flour bread: Optimization using D-optimal mixture design. *Heliyon* **2022**, *8*, e09514. [CrossRef] [PubMed]
24. Guo, S.; Wu, H.; Liu, X.; Zhao, W.; Zheng, J.; Li, W. Structural, Physicochemical and Digestive Property Changes of Potato Starch after Continuous and Repeated Dry Heat Modification and Its Comparative Study. *Foods* **2023**, *12*, 335. [CrossRef]
25. Ghalambor, P.; Asadi, G.; Mohammadi Nafchi, A.; Seyedin Ardebili, S.M. Investigation of dual modification on physicochemical, morphological, thermal, pasting, and retrogradation characteristics of sago starch. *Food Sci. Nutr.* **2022**, *10*, 2285–2299. [CrossRef] [PubMed]
26. Zhang, B.; Chen, M.; Xia, B.; Lu, Z.; Khoo, K.S.; Show, P.L.; Lu, F. Characterization and Preliminary Application of a Novel Lipoxygenase from *Enterovibrio norvegicus*. *Foods* **2022**, *11*, 2864. [CrossRef]
27. Huang, C.; Huang, J.; Zhang, B.; Omedi, J.O.; Chen, C.; Zhou, L.; Liang, L.; Zou, Q.; Zheng, J.; Zeng, Y.; et al. Rheo-Fermentation Dough Properties, Bread-Making Quality and Aroma Characteristics of Red Bean (*Vigna angularis*) Sourdough Induced by LAB Weissella confusa QS813 Strain Fermentation. *Foods* **2023**, *12*, 605. [CrossRef]
28. Kayama, K.; Wei, R.; Zhang, Y.; Wu, F.; Su, Z.; Dong, J.; Liu, X. Effects of Tea Powder on the Cooking Properties, Antioxidative Potential and Volatile Profiles of Dried Noodles. *Foods* **2022**, *11*, 858. [CrossRef]
29. Noonim, P.; Rajasekaran, B.; Venkatachalam, K. Effect of Palm Oil-Carnauba Wax Oleogel That Processed with Ultrasonication on the Physicochemical Properties of Salted Duck Egg White Fortified Instant Noodles. *Gels* **2022**, *8*, 487. [CrossRef]
30. Wang, R.; Li, M.; Wei, Y.; Guo, B.; Brennan, M.; Brennan, C.S. Quality Differences between Fresh and Dried Buckwheat Noodles Associated with Water Status and Inner Structure. *Foods* **2021**, *10*, 187. [CrossRef]
31. Yeoh, S.Y.; Tan, H.L.; Muhammad, L.; Tan, T.C.; Murad, M.; Mat Easa, A. Sensory, structural breakdown, microstructure, salt release properties, and shelf life of salt-coated air-dried yellow alkaline noodles. *NPJ Sci. Food* **2023**, *7*, 8. [CrossRef]
32. Hou, F.; Song, S.; Cui, W.; Yu, Z.; Gong, Z.; Wang, Y.; Wang, W. Flavor Improvement of Maillard Reaction Intermediates Derived from Enzymatic Hydrolysates of *Oudemansiella raphanipes* Mushroom. *Foods* **2024**, *13*, 1688. [CrossRef]
33. Zhou, S.; Feng, D.; Zhou, Y.; Duan, H.; He, Y.; Jiang, Y.; Yan, W. Characteristic Volatile Organic Compound Analysis of Different Cistanches Based on HS-GC-IMS. *Molecules* **2022**, *27*, 6789. [CrossRef] [PubMed]
34. Guardianelli, L.M.; Salinas, M.V.; Brites, C.; Puppo, M.C. Germination of White and Red Quinoa Seeds: Improvement of Nutritional and Functional Quality of Flours. *Foods* **2022**, *11*, 3272. [CrossRef] [PubMed]
35. Chipón, J.; Ramírez, K.; Morales, J.; Díaz-Calderón, P. Rheological and Thermal Study about the Gelatinization of Different Starches (Potato, Wheat and Waxy) in Blend with Cellulose Nanocrystals. *Polymers* **2022**, *14*, 1560. [CrossRef] [PubMed]
36. Wang, H.; Qiu, J.; Wu, Y.; Ouyang, J. Impact of soluble soybean polysaccharide on the gelatinization and retrogradation of corn starches with different amylose content. *Food Res. Int.* **2024**, *184*, 114254. [CrossRef] [PubMed]
37. Zhang, L.; Chen, J.; Xu, F.; Han, R.; Quan, M. Effect of Tremella fuciformis and Different Hydrocolloids on the Quality Characteristics of Wheat Noodles. *Foods* **2022**, *11*, 2617. [CrossRef]
38. Liu, Y.; Wei, Z.; Wang, J.; Wu, Y.; Xu, X.; Wang, B.; Abd El-Aty, A.M. Effects of different proportions of erythritol and mannitol on the physicochemical properties of corn starch films prepared via the flow elongation method. *Food Chem.* **2024**, *437*, 137899. [CrossRef]
39. Boonkor, P.; Sagis, L.M.C.; Lumdubwong, N. Pasting and Rheological Properties of Starch Paste/Gels in a Sugar-Acid System. *Foods* **2022**, *11*, 4060. [CrossRef]
40. Li, L.; Zhou, W.; Wu, A.; Qian, X.; Xie, L.; Zhou, X.; Zhang, L. Effect of Ginkgo Biloba Powder on the Physicochemical Properties and Quality Characteristics of Wheat Dough and Fresh Wet Noodles. *Foods* **2022**, *11*, 698. [CrossRef]

41. Li, S.; Chen, W.; Zongo, A.W.S.; Chen, Y.; Liang, H.; Li, J.; Li, B. Effects of non-starch polysaccharide on starch gelatinization and digestibility: A review. *Food Innov. Adv.* **2023**, *2*, 302–312. [CrossRef]
42. Wang, J.; He, Y.; Li, X.; Xie, Y.; Wang, X.; Zhu, D.; Liu, H. Effect of soluble soybean polysaccharides on the short- and long-term retrogradation properties of instant rice. *J. Sci. Food Agric.* **2023**, *103*, 4850–4857. [CrossRef] [PubMed]
43. Cui, C.; Caporaso, N.; Chen, J.; Fearn, T. Farinograph characteristics of wheat flour predicted by near infrared spectroscopy with an ensemble modelling method. *J. Food Eng.* **2023**, *359*, 111689. [CrossRef]
44. Dufour, M.; Chaunier, L.; Lourdin, D.; Réguerre, A.L.; Hugon, F.; Dugué, A.; Kansou, K.; Saulnier, L.; Della Valle, G. Unravelling the relationships between wheat dough extensional properties, gluten network and water distribution. *Food Hydrocoll.* **2024**, *146*, 109214. [CrossRef]
45. Liu, M.; Chen, G.; Zhang, H.; Yu, Q.; Mei, X.; Kan, J. Heat-induced inulin-gluten gel: Insights into the influences of inulin molecular weight on the rheological and structural properties of gluten gel to molecular and physicochemical characteristics. *Food Hydrocoll.* **2021**, *111*, 106397. [CrossRef]
46. Liu, S.; Liu, Q.; Li, X.; Obadi, M.; Jiang, S.; Li, S.; Xu, B. Effects of dough resting time on the development of gluten network in different sheeting directions and the textural properties of noodle dough. *LWT* **2021**, *141*, 110920. [CrossRef]
47. Okami, Y.; Tsunoda, H.; Watanabe, J.; Kataoka, Y. Efficacy of a meal sequence in patients with type 2 diabetes: A systematic review and meta-analysis. *BMJ Open Diabetes Res. Care* **2022**, *10*, 002534. [CrossRef]
48. Li, Q.M.; Li, Y.; Zou, J.H.; Guo, S.Y.; Wang, F.; Yu, P.; Su, X.J. Influence of Adding Chinese Yam (*Dioscorea opposita* Thunb.) Flour on Dough Rheology, Gluten Structure, Baking Performance, and Antioxidant Properties of Bread. *Foods* **2020**, *9*, 256. [CrossRef]
49. Shang, J.; Zhao, B.; Liu, C.; Li, L.; Hong, J.; Liu, M.; Zhang, X.; Lei, Y.; Zheng, X. Impact of wheat starch granule size on viscoelastic behaviors of noodle dough sheet and the underlying mechanism. *Food Hydrocoll.* **2023**, *134*, 108111. [CrossRef]
50. Sun, H.; Zhang, Y.; Sun, J. Dietary inulin supplementation improves the physicochemical and gel properties of duck myofibrillar protein: Insights into the effect of muscle fiber types. *Food Hydrocoll.* **2024**, *150*, 109722. [CrossRef]
51. Zhang, M.; Suo, W.; Deng, Y.; Jiang, L.; Qi, M.; Liu, Y.; Li, L.; Wang, C.; Zheng, H.; Li, H. Effect of ultrasound-assisted dough fermentation on the quality of dough and steamed bread with 50% sweet potato pulp. *Ultrason. Sonochem.* **2022**, *82*, 105912. [CrossRef] [PubMed]
52. Fan, H.; Fu, F.; Chen, Y.; Liu, M.; Ai, Z.; Bian, K. Effect of NaCl on rheological properties of dough and noodle quality. *J. Cereal Sci.* **2020**, *93*, 102936. [CrossRef]
53. Tang, P.; Zhang, S.; Meng, L.; Wang, Z.; Yang, Y.; Shen, X.; Tang, X. Effects of different content of EGCG or caffeic acid addition on the structure, cooking, antioxidant characteristics and in vitro starch digestibility of extruded buckwheat noodles. *Int. J. Biol. Macromol.* **2023**, *252*, 126426. [CrossRef] [PubMed]
54. Zhang, X.; Wang, A.; Yao, H.; Zhou, W.; Wang, M.; Liang, B.; Wang, F.; Tong, L.-T. Research advancements on the flavor compounds formation mechanism of pickled bamboo shoots in river snails rice noodles. *LWT* **2023**, *186*, 115226. [CrossRef]

foods

MDPI

Article

Genome Sequencing of Three Pathogenic Fungi Provides Insights into the Evolution and Pathogenic Mechanisms of the Cobweb Disease on Cultivated Mushrooms

Yufei Lan [1], Qianqian Cong [1], Qingwei Yu [1], Lin Liu [2], Xiao Cui [1], Xiumei Li [1], Qiao Wang [1], Shuting Yang [2], Hao Yu [2,*] and Yi Kong [1,*]

[1] Institute of Edible Fungi, Tai'an Academy of Agricultural Sciences, Tai'an 271000, China; lanyufei526@163.com (Y.L.); cong.qianqian@163.com (Q.C.); zsztyqw@126.com (Q.Y.); cuixiao916@163.com (X.C.); tanlixiumei@163.com (X.L.); 18815380509@163.com (Q.W.)

[2] Shandong Provincial Key Laboratory of Applied Mycology, School of Life Sciences, Qingdao Agricultural University, Qingdao 266109, China; liulin@qau.edu.cn (L.L.); yangshuting0701@163.com (S.Y.)

* Correspondence: yuhaosunshine@163.com (H.Y.); d5110@126.com (Y.K.)

Abstract: Fungal diseases not only reduce the yield of edible mushrooms but also pose potential threats to the preservation and quality of harvested mushrooms. Cobweb disease, caused primarily by fungal pathogens from the Hypocreaceae family, is one of the most significant diseases affecting edible mushrooms. Deciphering the genomes of these pathogens will help unravel the molecular basis of their evolution and identify genes responsible for pathogenicity. Here, we present high-quality genome sequences of three cobweb disease fungi: *Hypomyces aurantius* Cb-Fv, *Cladobotryum mycophilum* CB-Ab, and *Cladobotryum protrusum* CB-Mi, isolated from *Flammulina velutipes*, *Agaricus bisporus*, and *Morchella importuna*, respectively. The assembled genomes of *H. aurantius*, *C. mycophilum*, and *C. protrusum* are 33.19 Mb, 39.83 Mb, and 38.10 Mb, respectively. This is the first report of the genome of *H. aurantius*. Phylogenetic analysis revealed that cobweb disease pathogens are closely related and diverged approximately 17.51 million years ago. CAZymes (mainly chitinases, glucan endo-1,3-beta-glucosidases, and secondary metabolite synthases), proteases, KP3 killer proteins, lipases, and hydrophobins were found to be conserved and strongly associated with pathogenicity, virulence, and adaptation in the three cobweb pathogens. This study provides insights into the genome structure, genome organization, and pathogenicity of these three cobweb disease fungi, which will be a valuable resource for comparative genomics studies of cobweb pathogens and will help control this disease, thereby enhancing mushroom quality.

Keywords: edible mushrooms; Hypocreaceae; comparative genome; pathogenesis

1. Introduction

With the development of modern society and the increasing pursuit of healthful foods, edible mushrooms have become increasingly popular [1,2]. The total production of mushrooms and truffles worldwide reached 48.34 million metric tons in 2022 according to the data from the Statista website (statista.com). Most edible fungi are produced through commercially artificial cultivation [2–4]. Consequently, the outbreaks of diseases caused by fungal pathogens have continuously increased in recent years [5–7]. Mycopathogens cause heavy losses in commercial mushroom cultivation worldwide, such as cobweb disease, dry bubble disease, wet bubble disease, and green mold disease [6]. Cobweb disease is a significant limiting factor in edible mushroom production [8,9]. The main symptom of cobweb disease is a cobweb-like growth of fungal mycelia over the mushroom surface. The fruit body shows discoloration and rot and eventually becomes unsellable. The spores of pathogenic fungi are easily dispersed through air conditioning systems or by the action of watering [9]. If left untreated, the disease will spread throughout the mushroom crop through airborne spore dispersal [10]. In recent years, cobweb disease has been widespread and has caused serious losses in Europe, America, Africa, and Asia [9]. In the mid-1990s, cobweb disease emerged as the most serious disease affecting mushroom cultivation in the UK and Ireland. The prevalence of cobweb disease in *Agaricus bisporus* production has been documented at 33% in Turkey and 32% in Spain. Production losses can reach as high as 40% [9]. The earliest case of edible mushroom cobweb disease was reported in *A. bisporus* [11]. Subsequently, there has been an increasing number of reports describing the infection of other mushrooms by cobweb disease. Due to the inability to thoroughly sterilize the soil, mushrooms requiring casing soil in cultivation, such as *Coprinus comatus*, *Morchella importuna*, and *Oudemansiella raphanipes*, are particularly susceptible to cobweb disease [12–14]. A strain causing cobweb disease was isolated by Xu et al. and was from the most widely consumed mushroom *Lentinula edodes* [15]. Cobweb disease has also been observed in the widely cultivated *Pleurotus* spp., such as *P. ostreatus*, *P. pulmonarius*, and *P. eryngii* [16,17]. Industrially cultivated edible mushrooms, such as *Flammulina velutipes* and *Hypsizygus marmoreus*, could also be infected by cobweb mycopathogens [18,19]. Cobweb disease has become a serious disease in mushroom crops. Therefore, an in-depth study of the pathogenic fungi causing cobweb disease is imperative to better control this disease.

Strains from *Hypomyces* and *Cladobotryum* genera were the most commonly reported pathogenic fungi causing cobweb disease [8]. Both genera belong to the family Hypocreaceae, Ascomycota Fungi, indicating that the fungi responsible for cobweb disease are closely related, and that the pathogenic traits are inherited from ancestors. *Cladobotryum mycophilum*, the anamorph of *Hypomyces odoratus*, is currently the most cited causal agent of cobweb disease. It has been described as a pathogen of various edible mushrooms, including *A. bisporus*, *P. eryngii*, *G. lucidum*, and *Morchella sextelata* [13,20–22]. *Cladobotryum dendroides*, *Cladobotryum protrusum*, and *Cladobotryum varium* were also reported as infecting various species of edible mushrooms, such as *C. comatus*, *M. importuna*, *L. edodes*, *O. raphanipes*, *F. velutipes*, and *H. marmoreus* [14,15,19,23,24]. *Hypomyces*, another genus of pathogenic fungi causing cobweb disease, has been discovered infecting *F. velutipes*, *A. bisporus*, *Auricularia cornea*, and *Auricularia heimuer* with the corresponding pathogens *Hypomyces aurantius*, *Hypomyces mycophilus*, *Hypomyces cornea*, and *Hypomyces rosellus* [11,18,25–27]. Over the past years, studies on edible fungal pathogens have mainly focused on the identification of pathogens and fungicide resistance, while research on the pathogenic mechanism of the pathogen is rare. Genome sequencing and analysis have facilitated the study of the characteristics of edible fungal pathogens and the pathogenicity-related genes, laying a theoretical foundation and indicating the direction for the further systematic study of pathogenic mechanisms [28–30].

Recently, the genomes of a few parasitic fungi that cause dry bubble, wet bubble, and green mold diseases in edible mushrooms have been released [31–33]. *Cladobotryum protrusum* is the first fungus genome to be sequenced as the pathogenic fungus causing cobweb disease [24]. The analysis confirmed that the fungus belongs to the family Hypocreaceae, and genes from CAZymes, secondary metabolites, P450, and the pathogen–host interaction database (PHI) all contribute to its mycotrophic lifestyle. In subsequent studies, the genomes of two additional cobweb disease pathogenic fungi, *C. dendroides* and *C. mycophilum*, were sequenced [13,15]. Through genome analysis, one can examine genes that play an important role during the parasitic relationship with mushrooms, such as carbohydrate-active enzymes (CAZymes), cytochrome P450, peptidases, transporters, and genes involved in secondary metabolites. Understanding the pathogens at the molecular level is essential to gain new insights into the mechanisms of disease establishment within the host. To this aim, additional genomic studies are necessary, encompassing a broader array of mushroom fungal pathogens.

In the present study, three cobweb pathogenic fungi from different species were isolated from *A. bisporus*, *F. velutipes*, and *M. importuna*, respectively. The genome of the three fungal strains were sequenced and annotated. The objective of the study was to identify key factors causing cobweb disease through genomic and comparative genomic analysis. The analysis of these strains is also critical for studying and preventing cobweb disease.

2. Materials and Methods

2.1. Fungal Strains and Genomic DNA Preparation

The three strains, *C. protrusum*, *C. mycophilum*, and *H. aurantius*, used in this study were provided by Tai'an Academy of Agricultural Sciences (Tai'an, China). These strains were identified as the pathogens of cobweb disease that infect fruiting bodies of *M. importuna*, *A. bisporus*, and *F. velutipes*, respectively [18,23,34]. The mycelia of these strains were cultured and maintained on potato dextrose agar (PDA) plates. For genome sequencing, the mycelia were inoculated on the PDA plates overlaid with a cellophane membrane and cultured for 7 days at 25 °C in darkness. Genomic DNAs were extracted using the NucleoBond HMW DNA kit (Macherey–Nagel, Düren, Germany) according to the manufacturer's instructions. DNA concentration and purification were measured with the Nanodrop One (Thermo Fisher Scientific, Waltham, MA, USA).

2.2. Genome Sequencing and Assembly

The extracted genome DNA was sequenced using the technology of Oxford Nanopore and Illumina platforms. The Nanopore library was constructed using the Oxford Nanopore LSK-109 kit (Oxford Nanopore Technologies, Oxford, UK), and the Nanopore library was sequenced on the PromethION platform by Benagen Ltd. (Wuhan, China). After trimming and filtering, in total, clean data of 9.19, 10.75, and 8.18 Gb were generated for *C. protrusum*, *C. mycophilum*, and *H. aurantius*, respectively. Illumina paired-end sequencing was performed on an Illumina HiSeq2500 platform under 150 bp mode in the same company. The raw fastq data were filtered using the fastp software v0.23.2 [35]. After filtration, a total of 6.40, 6.26, and 6.55 Gb of clean data were kept for *C. protrusum*, *C. mycophilum*, and *H. aurantius*, respectively. Kmer (19 bp) was calculated using Jellyfish software v2.2.10, and the genome survey was analyzed using GenomeScope 2.0 (p = 1 and p = 2, m 20,000) [36]. The genome was assembled using Necat software v2020-08-03 with default parameters (GENOME_SIZE = 40,000,000, PREP_OUTPUT_COVERAGE = 100, and CNS_OUTPUT_COVERAGE = 90) [37]. Polishing of the assembled genome was performed in two iterations using Racon v1.5.0 (https://github.com/lbcb-sci/racon; accessed on 12 December 2023) with Nanopore reads and default parameters. The polished scaffolds were further polished with Pilon v1.24 through two iterations using filtered Illumina reads [38].

2.3. Gene Prediction and Function Annotation

Augustus software v3.4.0 was used to predict genes with *laccaria_bicolor* models, and *ab initio*-based gene prediction was performed using GeneMark-ES v4.69 with default parameters [39,40]. The two prediction results were integrated using EVidenceModeler software v1.1.1 [41]. The completeness of the gene prediction was evaluated using the BUSCO software v5.1.2 with fungi_odb10. Functional annotation of the three fungi's predicted protein-coding sequences (CDS) was performed by using Diamond to search against several protein databases such as the eggNOG-mapper, the Pfam database, and the SwissProt database [42,43]. Circular layouts were generated using Circos software v0.69 (http://circos.ca/; accessed on 20 December 2022) [44].

2.4. Comparative Genomic Analysis

Pairwise average nucleotide identity (ANI) was calculated using fastANI software v1.33 [45]. Collinearity analysis of pathogenic fungi (*H. aurantius*, *H. rosellus*, *C. mycophilum*, and *C. protrusum*) causing cobweb disease was conducted using MCScanX software [46]. Gene families and orthogroups were analyzed using the OrthoFinder software v2.5.4 [47]. The phylogenomic tree was constructed by concatenating single-copy orthologous protein sequences, which were then visualized using the FastTree software v2.1 [48]. Divergent time analysis was performed using PAML software packages v4 MCMCtree using reference divergence time between *Lentinula edodes* and *Saccharomyces cerevisiae* (626–806 MYA) and between *Aspergillus niger* and *Neurospora crassa* (233.8–367.0 MYA) obtained from the Timetree database http://timetree.org (accessed on 10 June 2024).

2.5. Analysis of Secretory Protein and Pathogenicity-Related Genes

CAZymes were annotated using dbCAN v3.0.2 software with the Hmmer search engine and default parameters [49]. Protein-signal peptides were predicted using SignalP v5.0b software, and transmembrane structures were predicted using TMHM v2.0c [50,51]. Signal-peptide proteins without transmembrane structures are considered secretory proteins. The proteins from PHI were downloaded from https://www.phi-base.org Version 4.17 (released on 1 May 2024; accessed on 20 June 2024) [52]. Proteins from the database of fungal virulence factors (DFVF) were downloaded from http://sysbio.unl.edu/DFVF (released in 2019, with 2058 entries; accessed on 20 June 2024) [53]. Genes in PHI and DFVF were identified using BlastP software v2.2.28 against these databases with an E-value of 0.00001.

3. Results and Discussion

3.1. Genome Information of Three Cobweb Disease Pathogens

Three mycoparasites causing cobweb disease were isolated and identified previously. *Hypomyces aurantius* CB-Fv was isolated from *F. velutipes* [18], *C. mycophilum* CB-Ab was isolated from a white button mushroom (*A. bisporus*) [34], and *C. protrusum* CB-Mi was isolated from cultivated *M. importuna* [23] (Figure 1).

A B C

Figure 1. Infected mushrooms and colony morphology of three cobweb disease pathogens used in this study. (**A**) Field symptoms of cobweb disease on *F. velutipes* and colonies of the cobweb disease strain *H. aurantius* CB-Fv on PDA media at 25 °C. (**B**) Field symptoms of cobweb disease on *A. bisporus* and colonies of the cobweb disease strain *C. mycophilum* CB-Ab. (**C**) Field symptoms of cobweb disease on *M. importuna* and colonies of the cobweb disease strain *C. protrusum* CB-Mi.

The pathogenicity of the three strains for cobweb disease was confirmed according to Koch's postulate previously. Here, the genome sequences of the three strains were sequenced using Oxford Nanopore and Illumina sequencing platforms. As shown in Supplementary Figure S1, a genome survey by GenomeScope was used to evaluate the heterozygosity of the three strains. The heterozygous rate was 0.073%, 0.128%, and 0.101% for *H. aurantius*, *C. mycophilum*, and *C. protrusum*, respectively. These results confirmed the homokaryotic nature of the three strains. The de novo assembly of the *H. aurantius* genome yielded 33.19 Mb, consisting of 10 contigs with a contig N50 length of 5.18 Mb and a contig N90 length of 2.99 Mb. The genome size of *C. mycophilum* was 39.83 Mb, consisting of 10 contigs with a contig N50 length of 5.50 Mb and a contig N90 length of 4.24 Mb. The *C. protrusum* genome contains a total length of 38.10 Mb, containing 10 contigs with a contig N50 length of 5.55 Mb and a contig N90 length of 2.41 Mb (Table 1, Figure 2). The genomes of the two *Cladobotryum* pathogens sequenced in this study exhibited similar genome sizes and GC content compared to those previously reported for the *Cladobotryum* species (Supplementary Table S1). This also suggested the completeness of the genomes sequenced in this study. The genome size of *H. aurantius* was smaller than that of other reported cobweb disease pathogens from the genus *Cladobotryum* (>36 Mb), and it was also smaller than the genomes of the other two sequenced strains from *Hypomyces* genus (>38.48 Mb) (Supplementary Table S1). In addition, the GC content of the *H. aurantius* genome was the highest among these reported strains. *Hypomyces aurantius* represents the first cobweb pathogen from the *Hypomyces* genus to be sequenced, providing valuable information for gaining deeper insights into cobweb disease pathogens. Most fungal diseases are attributed to filamentous fungi, primarily due to the similarity in their nutritional requirements to those of edible mushrooms. These fungi can secrete numerous proteins, allowing them to parasitize edible mushrooms by outcompeting for nutrients and through direct invasion. The genomes of these fungi generally range from 35 to 45 Mb, likely owing to the requirement to encode an extensive array of CAZymes and other pathogenic genes [54–57]. In contrast, yeast-like pathogenic fungi, such as *Saccharomycopsis phalluae*, possess a genome size of merely 14 Mb. Analyses of published filamentous fungal genomes indicate that various pathogenic fungi exhibit similar potential pathogenic mechanisms [58].

Table 1. De novo genome assembly and features of *H. aurantius*, *C. mycophilum*, and *C. protrusum*.

	***H. aurantius* CB-Fv**	***C. mycophilum* CB-Ab**	***C. protrusum* CB-Mi**
Genome size	33.19 Mb	39.83 Mb	38.10 Mb
Total contig number	10	10	11
Longest contig	6.38 Mb	7.44 Mb	7.53 Mb
Contig N50	5.18 Mb	5.50 Mb	5.55 Mb
Genome size	33.19 Mb	39.83 Mb	38.10 Mb
Contig N90	2.99 Mb	4.24 Mb	2.41 Mb
GC content	49.00%	48.00%	48.00%
Sequencing platform	Napopore, Illumina	Napopore, Illumina	Napopore, Illumina
Isolation host	*F. velutipes*	*A. bisporus*	*M. importuna*

3.2. Genome Annotation

The protein-encoding genes in the assembled genomic sequence were predicted based on sequence homology and de novo gene predictions. Gene models of 9791, 11,365, and 113,038 were predicted with a total length of 15.66 Mbp (47.2% of the genome), 18.11 Mbp (45.5% of the genome), and 17.86 Mbp (46.9% of the genome) for *H. aurantius*, *C. mycophilum*, and *C. protrusum* (Figure 2, Table 2). The predicted genes in *H. aurantius*, *C. mycophilum*, and *C. protrusum* genomes showed an average length of 1590.3 bp, 1593.5 bp, and 1618.1 bp, an average exon length of 554.7 bp, 561.1 bp, and 572.9 bp, an average intron length of 83.6 bp, 88.5 bp, and 83.1 bp, and an average intron number of 1.87, 1.84, and 1.82, respectively (Table 2). The completeness of assemblies was evaluated using BUSCO v5.1.2 (fungi_odb10, ascomycota_odb10, sordariomycetes_odb10) [59]. The results of BUSCO analysis of *H. aurantius* CB-Fv, *C. mycophilum* CB-Ab, and *C. protrusum* CB-Mi based on sordariomycetes_odb10 were 98.2%, 98.4%, and 98.8%, respectively. The high BUSCO completeness indicated the quality of the genome assembly and high fidelity of gene prediction.

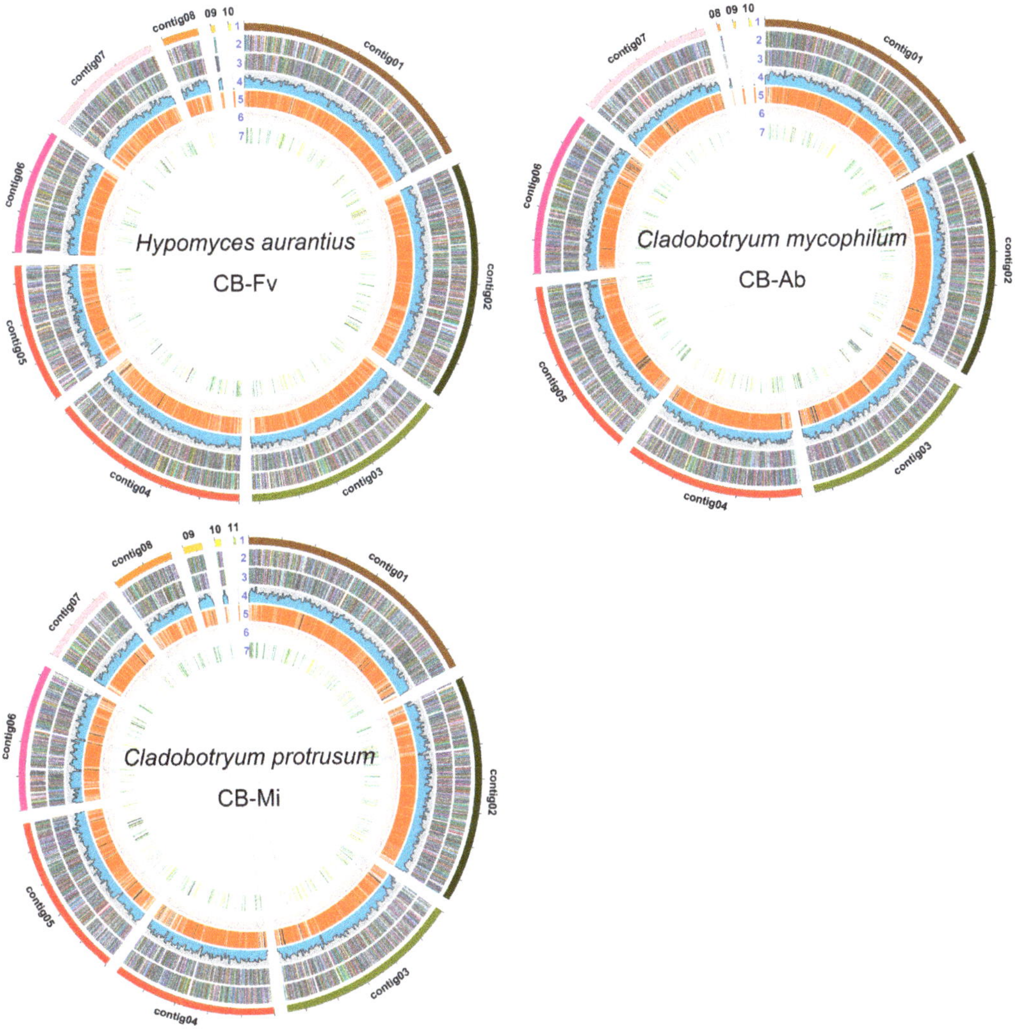

Figure 2. Genome maps of three cobweb disease pathogens *H. aurantius* CB-Fv, *C. mycophilum* CB-Ab, and *C. protrusum* CB-Mi. Layer 1, contigs; Layers 2 and 3, predicted genes in forward and reverse strand; Layer 4, gene density; Layer 5, repeat sequences; Layer 6, GC content; Layer 7, CAZymes. Links within and between contigs represent collinear blocks generated from MCScanX. The plot was visualized using Circos software with a window size of 50 kb.

Table 2. Characteristics of the gene prediction of *H. aurantius*, *C. mycophilum*, and *C. protrusum*.

	H. aurantius CB-Fv	*C. mycophilum* CB-Ab	*C. protrusum* CB-Mi
Gene number	9791	11,365	11,038
Total length	15.57 Mb	18.11 Mb	17.86 Mb
Average length	1590.3 bp	1593.1 bp	1618.1 bp
Average exon length	554.7 bp	561.1 bp	572.9 bp
Average exon number	2.87	2.84	2.82
Average Intron length	83.6 bp	88.5 bp	83.1 bp
Average intron number	1.87	1.84	1.82

Functional annotations of the predicted genes were performed using publicly available databases of Swissprot, GO, KEGG, eggNOG, and Pfam. Of the identified genes, the largest number of genes was annotated by the eggNOG database, and the least number of genes was annotated by GO analysis (Supplementary Table S2). There were 5843, 6580, and 6511 genes of *H. aurantius*, *C. mycophilum*, and *C. protrusum* annotated in the SwissProt database, accounting for 59.68%, 57.90%, and 58.99% of the total number of genes. Based on the similarity of protein domains, 7278 (74.33%), 8390 (73.82%), and 8214 (74.42%) genes of *H. aurantius*, *C. mycophilum*, and *C. protrusum* were annotated by Pfam database. There were 3951 (40.35%), 4148 (36.50%), and 4092 (37.07%) genes of *H. aurantius*, *C. mycophilum*, and *C. protrusum* annotated by the KEGG pathway, of which most genes involved in metabolism (Supplementary Table S2).

3.3. Evolution and Comparative Genomic Analysis

Both *Cladobotryum* and *Hypomyces* come from the family Hypocreaceae. Genomes of seven genera, including *Cladobotryum* and *Hypomyces*, of the Hypocreaceae family have been published in the NCBI database (May 2024). The comparison of all genomes from *Cladobotryum* and *Hypomyces* and the representative strains from the other five genera were performed using fastANI software. While the ANI value is typically employed to determine species boundaries of prokaryotic genomes [45], the software was simply used to present the genome similarities between different fungal genomes in this study. Figure 3 reveals that strains from the same species showed high ANI values (>98%) with each other. *Cladobotryum protrusum* strain CB-Mi shared the highest ANI value of 98.96% with *C. protrusum* strain CCMJ2080, also a cobweb pathogen isolated from *C. comatus*, while the ANI values with other fungi were all lower than 85%. Strain CB-Ab showed the highest ANI values of 98.49% and 98.95% with strain ATHUM6906 and strain CCMJ2923, respectively, while ANI values with other strains were all below 90%. Strain CB-Fv, the first sequenced strain of *H. aurantius*, exhibited its highest ANI value of 79.59% with *H. perniciosus* HP10. Thus, ANI analysis corroborates taxonomic classifications to some extent, reflecting the evolutionary relationships among different strains.

To further investigate the evolutionary relationships of the three fungi and the divergent time from a common ancestor, phylogenetic analyses were conducted between the three fungi and 24 other fungi species (20 Ascomycetes and 4 Basidiomycetes). A total of 101 single-copy orthologous genes were identified using OrthoFinder and used for phylogenetic reconstruction, and species divergence time was estimated based on amino-acids sequences using mcmctree (Figure 3A). The results indicated that strains from the Sordariomycetes class cluster together. The genera *Hypomyces* and *Escovopsis* diverged 17.51 million years ago. Based on ANI analysis and genomic phylogenetic analysis, strain CCMJ2808 should belong to the genus *Cladobotryum*. The genera *Hypomyces* and *Cladobotryum* separated 17.12 million years ago. This suggests that cobweb disease pathogens share genetically similar, indicating a common infection mechanism at the genetic level.

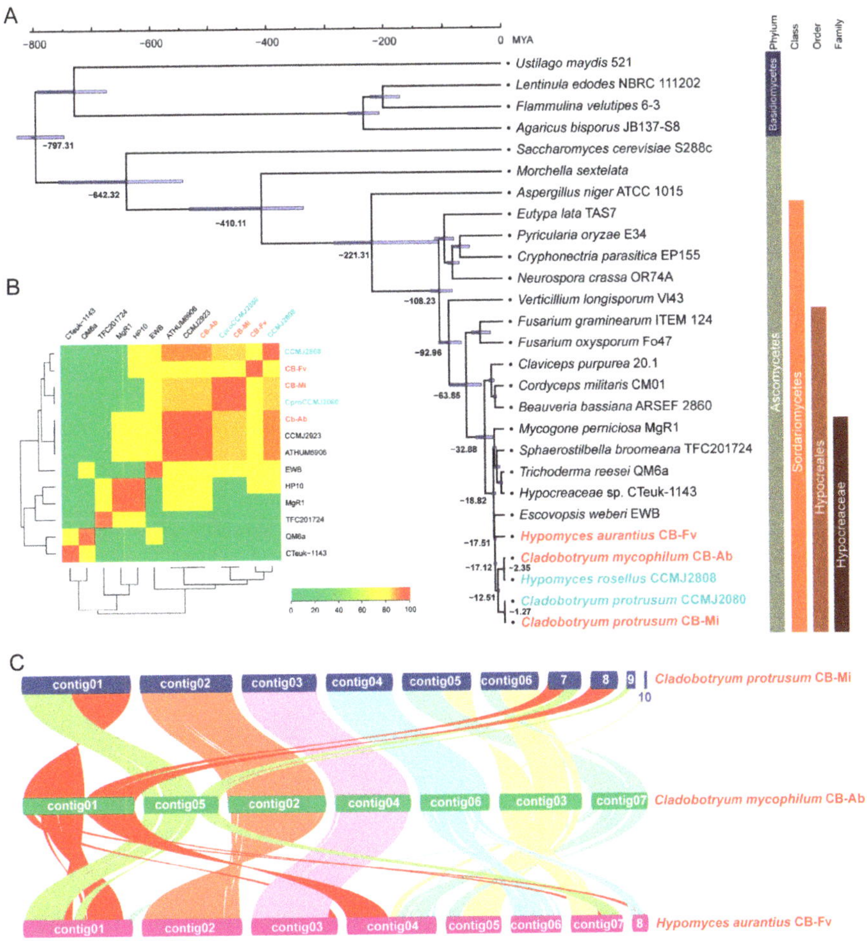

Figure 3. Evolutionary and comparative genomic analysis of three cobweb disease strains. (**A**) Fungi phylogenetic tree inferred from 101 single-copy orthologs among 27 fungi species identified using OrthoFrinder software. (**B**) The cluster heatmap of ANI values between 13 strains and the specific value were listed in Supplementary Table S3. (**C**) Genome collinearity among three cobweb disease strains, *H. aurantius* Cb-Fv, *C. mycophilum* CB-Ab, and *C. protrusum* CB-Mi. Each line connects a pair of collinearity blocks between two genomes.

Collinearity analysis of the three strains revealed a high degree of synteny conservation, as illustrated in Figure 3C: Contig02 and contig04 of strain CB-Ab were highly conserved, exhibiting almost no chromosomal recombination events among the three strains. Portions of contig01 and contig05 of strain CB-Ab displayed collinearity with contig01 of strains CB-Mi and contig01 of strain CB-Fv, whereas the remaining parts of the two contigs are dispersed across different locations in the other two strain genomes. Contig06 of strain CB-AB showed high collinearity with contig04 of strain CB-Mi but was fragmented across different positions in the genome of strain CB-Fv. Contigs03 and contig07 of strain CB-Ab exhibited numerous fragmentation events compared to the other two genomes. Overall, fewer genomic rearrangement events occurred between strains CB-Mi and CB-Ab compared to CB-Fv and the other two strains, which was consistent with the previous phylogenetic analysis. The results indicated that there are conserved contigs and variable contigs at the chromosomal level, with more rearrangement events occurring in the non-conserved variable contigs.

3.4. Comparative Analysis of CAZymes

The genome of *H. aurantius* CB-Fv, *C. mycophilum* CB-Afb, and *C. protrusum* CB-Mi contained 358, 399, and 401 CAZymes with a high diversity of families, including a total of six classes: glycoside hydrolases (GHs = 173, 184, and 197), glycosyl transferases (GTs = 82, 81, and 80), carbohydrate esterases (CEs = 17, 19, and 19), auxiliary activities (AAs = 72, 101, and 91), carbohydrate-binding molecules (CBMs = 9, 6, and 6), and polysaccharide lyases (PLs = 5, 8, and 8) (Figure 4A, Supplementary Tables S4–S6). The family with the highest number of genes was the AA7 family. The three strains, *H. aurantius* CB-Fv, *C. mycophilum* CB-Ab, and *C. protrusum* CB-Mi, had 36, 51, and 45 genes, respectively, in this family. Genes in this family are primarily involved in the synthesis of secondary metabolites, such as the antifungal substance trichoxide, the defense-related metabolite zearalenone, and chanoclavine [60,61]. Cell wall-degrading enzymes have been reported to be associated with mycoparasitism. The white mildew disease fungus *Calcarisporium cordycipiticola* could degrade the hyphase of its host *Cordyceps militaris* [57]. Genes from the GH18 family numbers were 27, 22, and 27 in the three pathogens, respectively, second only to the AA7 family (Figure 4B, Supplementary Tables S4–S6). The enzymes of this family were mainly annotated as endochitinases and chitotriosidases. These enzymes can hydrolyze the β-(1, 4)-linkage between adjacent N-acetyl glucosamine residues of chitin, playing a pivotal role in the context of infectious diseases [62]. The induction of chitinolytic enzymes were observed in the interaction between *Trichoderma harzianum* and *Sclerotium rolfsii* [63]. A transcriptome analysis of the wet bubble disease *Mycogone perniciosa* revealed that several GH18 chitinases were significantly upregulated during the infection process targeting *A. bisporus* [64]. Calonje et al. [65] reported that hydrolytic enzymes, including chitinases, were produced by *Verticillium fungicola*, a dry bubble pathogen, in cultures with *A. bisporus* vegetative mycelial cell walls as the only carbon source. Similar results were also reported for the green mould disease fungus *Trichoderma aggressivum* [66]. The results were consistent with our genomic analysis results, indicating that genes in the GH18 family play a crucial role in cobweb disease pathogenicity [67]. Another gene family, GH55, mainly consists of glucan endo-1,3-beta-glucosidase, which are also pathogenesis-related genes involved in the degradation of beta-1,3-glucan in host cell walls [68]. The enzyme activity of 1,3-beta-glucanase was also detected in the culture of *V. fungicola* with *A. bisporus* myvelial cell walls as the substrate [65]. The three pathogens had five, seven, and eight genes, respectively, in the GH55 family, which is also higher than the number in *F. velutipes* (3), *A. bisporus* (1), and *P. ostreatus* (2), suggesting a possible link to their pathogenicity. Overall, genomic analysis confirmed that cell wall-degrading enzymes play key roles in the infection of mushrooms.

3.5. Pathogenicity-Related Genes

As secreted proteins are critical virulence determinants in fungal pathogens, we compared the secreted proteins of three strains to gain information about their composition and conservation. A total of 717, 882, and 936 secreted proteins (with signal peptide and without transmembrane structure) were predicted in the genomes of strain CB-Fv, CB-Ab, and CB-Mi, respectively (Figure 5A, Supplementary Tables S4–S6). To identify potentially pathogenic genes in these cobweb disease fungi, the whole genomes were analyzed via BLASTP against the PHI and DFVF databases, which predict protein function during host infection. There were 2506, 2758, and 2736 genes annotated in the PHI database for *H. aurantius* CB-Fv, *C. mycophilum* CB-Ab, and *C. protrusum* CB-Mi, respectively, based on sequence identity ≧ 30% and alignment coverage ≥ 50% (Figure 5A, Supplementary Tables S4–S6). The DFVF database is an inclusive database of recognized fungal virulence factors. It collects 2058 pathogenic genes released by 228 fungal strains from 85 genera. Currently, there were 746, 846, and 829 genes identified in the genomes of *H. aurantius*, *C. mycophilum*, and *C. protrusum*, respectively (sequence identity $\geq$ 30% and alignment coverage $\geq$ 50%), against the DFVF database (Figure 5A, Supplementary Tables S4–S6). In the genomes of *H. aurantius* CB-Fv, *C. mycophilum* CB-Ab, and *C. protrusum* CB-Mi, 89, 109, and 112 secreted proteins (common genes) were identified in both the PHI and DFVF databases, respectively. Among the common genes, the majority of proteins (>90%) of each strain had homologs in the common genes of the other two strains, indicating that these proteins (core genes) were essential to the pathogenicity of the three web blight pathogens. Some of the core genes were CAZymes, including the previously mentioned chitinases and glucanases, while most of the other genes were annotated as proteases (peptidases), lipases, hydrophobins, KP4 killer proteins, and ROS-related enzymes (Figure 5B, Supplementary Tables S4–S6). The induction of a laccase gene, lcc2, by *A. bisporus* can provide defense against *T. aggressivum*, a green mould disease. The expression of proteases might selectively degrade the defensive enzymes secreted by the host mushroom [69]. The upregulated expression of a proteinase gene, *prb1*, was detected when *T. aggressivum* was co-cultured with *A. bisporus* [66]. The results suggested that these genes have roles in pathogenicity. Lu's study revealed that ROS levels increased when *Phallus rubrovolvatus* was co-cultured with a pathogenic fungus *Trichoderma koningiopsis* [70]. Therefore, ROS-related enzymes in cobweb disease may be associated with stress defense against host mushrooms.

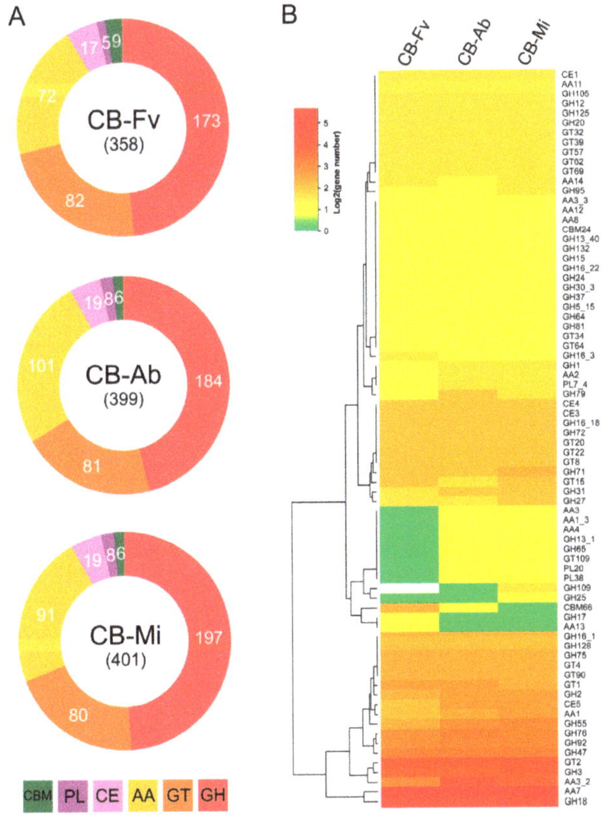

Figure 4. Distribution of CAZymes genes in three cobweb pathogens. (**A**) Distribution of CAZymes genes in *H. aurantius* Cb-Fv, *C. mycophilum* CB-Ab, and *C. protrusum* CB-Mi. (**B**) Heatmap representing the gene number (>1) of CAZyme families in three strains. Abbreviations: GH: glycoside hydrolase, GT: glycosyl transferases, CE: carbohydrate esterases, AA: auxiliary activities, CBM: carbohydrate-binding molecules, PL: polysaccharide lyases.

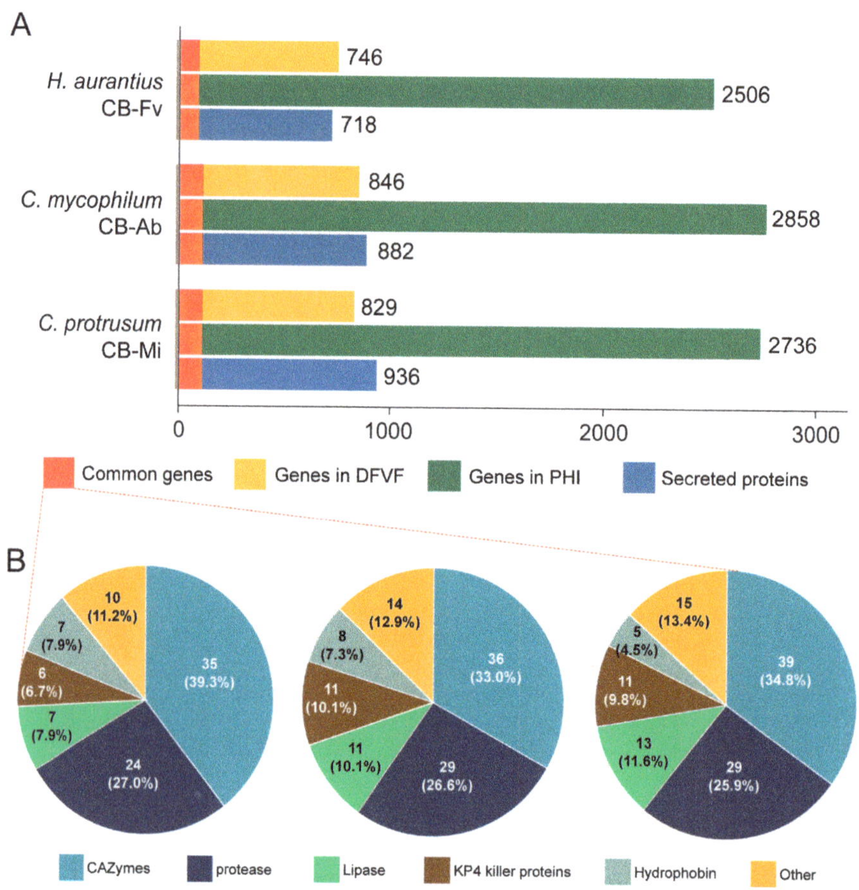

Figure 5. Distribution of pathogenicity-related genes in three cobweb disease strains. (**A**) The distribution of secreted proteins, genes identified in the PHI database, and genes identified in the DFVF database in *H. aurantius* Cb-Fv, *C. mycophilum* CB-Ab, and *C. protrusum* CB-Mi, respectively. Common genes were genes that belong to all these categories. (**B**) The gene distribution in common genes of the three cobweb disease strains.

Based on the analysis of genome, we can infer that the cobweb disease pathogens may inhibit the growth of host cells by secreting antifungal metabolites (compounds synthesized by enzymes from the AAs family and KP4 killer proteins [71]). Subsequently, they secrete chitinases, glucan endo-1,3-beta-glucosidases, and other enzymes (such as proteases and lipases) to degrade the host cell walls, providing nutrients for their own growth and, thus, completing the infection process. Additionally, hydrophobins are deeply involved in and regulate the infection processes. In future research, it will be essential to investigate the specific secreted proteins and CAZymes related to pathogenicity through transcriptomics or proteomics.

4. Conclusions

In summary, high-quality genomes of three cobweb disease fungi, *H. aurantius* Cb-Fv, *C. mycophilum* CB-Ab, and *C. protrusum* CB-Mi, were sequenced using the Illumina and Nanopore platforms. Genomic and comparative analysis indicate that the pathogens of cobweb disease might inhibit the growth of host mycelia by secreting antifungal substances and cell wall-degrading enzymes to hydrolyze the host mycelia, thereby completing the infection process. The in-depth genomic information generated in this study will contribute to the control of cobweb disease in mushroom production.

Supplementary Materials: The following supporting information can be downloaded at: https://www.mdpi.com/article/10.3390/foods13172779/s1, Figure S1: Genome survey results generated by Genome Scope software; Figure S2: BUSCO analysis results of three cobweb disease fungi; Table S1: Characteristics of genomes from *Hypomyces* and *Cladobotryum* genera; Table S2: Predicted gene numbers identified in eggNOG, SwissProt, Pfam, KEGG, and GO databases in three cobweb disease strains; Table S3: ANI values between the genome of different fungi. The ANI values were generated using fastANI software; Table S4: Pathogenicity-related genes in *Hypomyces aurantius* Cb-Fv; Table S5: Pathogenicity-related genes in *Cladobotryum mycophilum* CB-Ab; Table S6: Pathogenicity-related genes in *Cladobotryum protrusum* CB-Mi.

Author Contributions: Conceptualization, Y.L. and Y.K.; methodology, Y.L., H.Y., L.L. and Q.C.; formal analysis, H.Y. and S.Y.; investigation, Q.Y.; Resources, Y.L.; Software, X.C.; data curation, X.L. and Q.W.; writing—original draft preparation, Y.L. and L.L.; writing—review and editing, H.Y.; supervision, Y.K.; project administration, Y.K.; funding acquisition, Y.L. All authors have read and agreed to the published version of the manuscript.

Funding: This research was funded by Shandong Edible Fungus Agricultural Technology System (SDAIT-07-04, SDAIT-07-02).

Institutional Review Board Statement: Not applicable.

Informed Consent Statement: Not applicable.

Data Availability Statement: The original contributions presented in the study are included in the article/Supplementary Material, further inquiries can be directed to the corresponding authors or at https://file.mushroomlab.cn.

Conflicts of Interest: The authors declare no conflicts of interest.

Abbreviation

Agaricus bisporus (*A. bisporus*); *Coprinus comatus* (*C. comatus*); *Morchella importuna* (*M. importuna*); *Oudemansiella raphanipes* (*O. raphanipes*); *Lentinula edodes* (*L. edodes*); *Pleurotus ostreatus* (*P. ostreatus*); *Pleurotus pulmonarius* (*P. pulmonarius*); *Pleurotus eryngii* (*P. eryngii*); *Flammulina velutipes* (*F. velutipes*); *Hypsizygus marmoreus* (*H. marmoreus*); *Cladobotryum mycophilum* (*C. mycophilum*); *Hypomyces odoratus* (*H. odoratus*); *Morchella sextelata* (*M. sextelata*); *Cladobotryum dendroides* (*C. dendroides*); *Cladobotryum protrusum* (*C. protrusum*); *Cladobotryum varium* (*C. varium*); *Auricularia cornea* (*A. cornea*); *Auricularia heimuer* (*A. heimuer*); *Hypomyces aurantius* (*H. aurantius*); *Hypomyces mycophilus* (*H. mycophilus*); *Hypomyces cornea* (*H.s cornea*); *Hypomyces rosellus* (*H. rosellus*); *Trichoderma aggressivum* (*T. aggressivum*); carbohydrate-active enzymes (CAZymes); potato dextrose agar (PDA); coding sequences (CDSs); average nucleotide identity (ANI); pathogen-host interaction database (PHI); database of fungal virulence factors (DFVF); glycoside hydrolase (GH); glycosyl transferase (GT); carbohydrate esterase (CE); auxiliary activity (AA); carbohydrate-binding molecule (CBM); polysaccharide lyase (PL).

References

1. Bell, V.; Silva, C.R.P.G.; Guina, J.; Fernandes, T.H. Mushrooms as future generation healthy foods. *Front. Nutr.* **2022**, *9*, 1050099. [CrossRef] [PubMed]
2. Li, C.; Xu, S. Edible mushroom industry in China: Current state and perspectives. *Appl. Microbiol. Biotechnol.* **2022**, *106*, 3949–3955. [CrossRef] [PubMed]
3. Grimm, D.; Wösten, H.A.B. Mushroom cultivation in the circular economy. *Appl. Microbiol. Biotechnol.* **2018**, *102*, 7795–7803. [CrossRef] [PubMed]
4. Zhang, Y.; Mao, C.; Liu, X.; Guo, L.; Hu, C.; Li, X.; Xu, L.; Yu, H. Insights into the evolution and mechanisms of response to heat stress by whole genome sequencing and comparative proteomics analysis of the domesticated edible mushroom *Lepista sordida*. *Mycology* **2024**, 1–20. [CrossRef]
5. Kertesz, M.A.; Thai, M. Compost bacteria and fungi that influence growth and development of *Agaricus bisporus* and other commercial mushrooms. *Appl. Microbiol. Biotechnol.* **2018**, *102*, 1639–1650. [CrossRef]

6. Gea, F.J.; Navarro, M.J.; Santos, M.; Diánez, F.; Carrasco, J. Control of fungal diseases in mushroom crops while dealing with fungicide resistance: A review. *Microorganisms* **2021**, *9*, 585. [CrossRef]
7. Yang, X.; Li, S.; Li, X.; Zhang, C.; Liu, M.; Guo, L.; Liu, L.; Yu, H. Comparative proteomics reveals the response and adaptation mechanisms of white *Hypsizygus marmoreus* against the biological stress caused by *Penicillium*. *Food Sci. Hum. Well.* **2024**, *13*, 1645–1661. [CrossRef]
8. Tamm, H.; Põldmaa, K. Diversity, host associations, and phylogeography of temperate aurofusarin-producing *Hypomyces/Cladobotryum* including causal agents of cobweb disease of cultivated mushrooms. *Fungal Biol.* **2013**, *117*, 348–367. [CrossRef]
9. Carrasco, J.; Navarro, M.J.; Gea, F.J. Cobweb, a serious pathology in mushroom crops: A review. *Span. J. Agric. Res.* **2017**, *15*, e10R01. [CrossRef]
10. Adie, B.; Grogan, H.; Archer, S.; Mills, P. Temporal and spatial dispersal of *Cladobotryum conidia* in the controlled environment of a mushroom growing room. *Appl. Environ. Microb.* **2006**, *72*, 7212–7217. [CrossRef]
11. Fletcher, J.T.; Hims, M.J.; Hall, R.J. The control of bubble diseases and cobweb disease of mushrooms with prochloraz. *Plant Pathol.* **1983**, *32*, 123–131. [CrossRef]
12. Wang, G.Z.; Guo, M.P.; Bian, Y.B. First report of *Cladobotryum protrusum* causing cobweb disease on the edible mushroom *Coprinus comatus*. *Plant Dis.* **2015**, *99*, 287. [CrossRef] [PubMed]
13. Liu, Z.; Cong, Y.; Sossah, F.L.; Lu, Y.; Kang, J.; Li, Y. Characterization and genome analysis of *Cladobotryum mycophilum*, the causal agent of cobweb disease of *Morchella sextelata* in China. *J. Fungi* **2023**, *9*, 411. [CrossRef] [PubMed]
14. Qin, W.; Li, J.; Zeng, Z.; Wang, S.; Gao, L.; Rong, C.; Gao, Q.; Liu, Y. First report of cobweb disease in *Oudemansiella raphanipes* caused by *Cladobotryum varium* in Beijing, China. *Plant Dis.* **2021**, *105*, 4171. [CrossRef] [PubMed]
15. Xu, R.; Liu, X.; Peng, B.; Liu, P.; Li, Z.; Dai, Y.; Xiao, S. Genomic features of *Cladobotryum dendroides*, which causes cobweb disease in edible mushrooms, and identification of genes related to pathogenicity and mycoparasitism. *Pathogens* **2020**, *9*, 232. [CrossRef]
16. Kim, M.K.; Seuk, S.W.; Lee, Y.H.; Kim, H.R.; Cho, K.M. Fungicide sensitivity and characterization of cobweb disease on a *Pleurotus eryngii* mushroom crop caused by *Cladobotryum mycophilum*. *Plant Pathol. J.* **2014**, *30*, 82–89. [CrossRef] [PubMed]
17. Mignucci, J.S.; Hernández-Bacó, C.; Rivera-Vargas, L.; Betancourt, C.; Alameda, M. Diseases and pests research on oyster mushrooms (*Pleurotus* spp.) in Puerto Rico. *Int. J. Mushroom Sci.* **2000**, *3*, 21–26.
18. Lan, Y.F.; Cong, Q.Q.; Cui, X. Identification of the Pathogen Causing Cobweb Disease on the *Flammulina velutipes*. *Edible Fungi China* **2018**, *38*, 76–79. [CrossRef]
19. Back, C.G.; Lee, C.Y.; Seo, G.S.; Jung, H.Y. Characterization of species of *Cladobotryum* which cause cobweb disease in edible mushrooms grown in Korea. *Mycobiology* **2012**, *40*, 189–194. [CrossRef]
20. McKay, G.J.; Egan, D.; Morris, E.; Scott, C.; Brown, A.E. Genetic and morphological characterization of *Cladobotryum* species causing cobweb disease of mushrooms. *Appl. Environ. Microbiol.* **1999**, *65*, 606–610. [CrossRef]
21. Gea, F.J.; Navarro, M.J.; Suz, L.M. First report of *Cladobotryum mycophilum* causing cobweb on cultivated king oyster mushroom in Spain. *Plant Dis.* **2011**, *95*, 1030. [CrossRef] [PubMed]
22. Zuo, B.; Lu, B.H.; Liu, X.L.; Wang, Y.; Ma, G.L.; Gao, J. First report of *Cladobotryum mycophilum* causing cobweb on *Ganoderma lucidum* cultivated in Jilin province, China. *Plant Dis.* **2016**, *100*, 1239. [CrossRef]
23. Lan, Y.F.; Cong, Q.Q.; Wang, Q.W.; Tang, L.N.; Li, X.M.; Yu, Q.W.; Cui, X.; An, X.R.; Yu, C.X.; Kong, F.H.; et al. First report of *Cladobotryum protrusum* causing cobweb disease on cultivated *Morchella importuna*. *Plant Dis.* **2020**, *104*, 977. [CrossRef]
24. Sossah, F.L.; Liu, Z.; Yang, C.; Okorley, B.A.; Sun, L.; Fu, Y.; Li, Y. Genome sequencing of *Cladobotryum protrusum* provides insights into the evolution and pathogenic mechanisms of the cobweb disease pathogen on cultivated mushroom. *Genes* **2019**, *10*, 124. [CrossRef] [PubMed]
25. Xie, J.; Liu, X.; Lu, J.; Qin, Z.; Yu, N.; Zeng, X.Y.; Tian, F. First report of cobweb disease in *Auricularia cornea* caused by *Hypomyces mycophilus* in Guizhou, China. *Plant Dis.* **2023**, *108*, 204. [CrossRef]
26. Zhang, J.; Wen, D.; Zhu, Y.; Wen, Z. First report of cobweb disease of *Auricularia heimuer* caused by *Hypomyces mycophilus* in Fujian province, China. *Plant Dis.* **2023**, *107*, 2543. [CrossRef] [PubMed]
27. Xie, J.; Lu, S.; Tarafder, E.; Pan, Y.; Peng, K.; Zeng, X.; Tian, F. Taxonomy, biological characterization and fungicide sensitivity assays of *Hypomyces cornea* sp. nov. causing cobweb disease on *Auricularia cornea*. *Fungal Biol.* **2024**, *128*, 1616–1625. [CrossRef] [PubMed]
28. Möller, M.; Stukenbrock, E.H. Evolution and genome architecture in fungal plant pathogens. *Nat. Rev. Microbiol.* **2017**, *15*, 756–771. [CrossRef]
29. Sharma, K.K. Fungal genome sequencing: Basic biology to biotechnology. *Crit. Rev. Biotechnol.* **2016**, *36*, 743–759. [CrossRef]
30. Li, S.; Zhao, S.; Hu, C.; Mao, C.; Guo, L.; Yu, H.; Yu, H. Whole genome sequence of an edible mushroom *Stropharia rugosoannulata* (Daqiugaigu). *J. Fungi* **2022**, *8*, 99. [CrossRef]
31. Berendsen, R.L.; Baars, J.J.; Kalkhove, S.I.; Lugones, L.G.; Wösten, H.A.; Bakker, P.A. *Lecanicillium fungicola*: Causal agent of dry bubble disease in white-button mushroom. *Mol. Plant Pathol.* **2010**, *11*, 585–595. [CrossRef]
32. Wang, X.; Peng, J.; Sun, L.; Bonito, G.; Guo, Y.; Li, Y.; Fu, Y. Genome sequencing of *Paecilomyces penicillatus* provides insights into its phylogenetic placement and mycoparasitism mechanisms on morel mushrooms. *Pathogens* **2020**, *9*, 834. [CrossRef]
33. Li, D.; Sossah, F.L.; Sun, L.; Fu, Y.; Li, Y. Genome analysis of *Hypomyces perniciosus*, the causal agent of wet bubble disease of button mushroom (*Agaricus bisporus*). *Genes* **2019**, *10*, 417. [CrossRef] [PubMed]

34. Lan, Y.F.; Wang, Q.W.; Yu, C.X.; Cong, Q.Q.; Tang, L.N.; An, X.R.; Kong, F.H.; Li, X.D. First report of *Cladobotryum mycophilum* causing cobweb disease on *Agaricus bisporus* in China. *Plant Dis.* **2016**, *100*, 2334. [CrossRef]
35. Chen, S.; Zhou, Y.; Chen, Y.; Gu, J. fastp: An ultra-fast all-in-one FASTQ preprocessor. *Bioinformatics* **2018**, *34*, i884–i890. [CrossRef]
36. Ranallo-Benavidez, T.R.; Jaron, K.S.; Schatz, M.C. GenomeScope 2.0 and Smudgeplot for reference-free profiling of polyploid genomes. *Nat. Commun.* **2020**, *11*, 1432. [CrossRef]
37. Chen, Y.; Nie, F.; Xie, S.Q.; Zheng, Y.F.; Dai, Q.; Bray, T.; Wang, Y.X.; Xing, J.F.; Huang, Z.J.; Wang, D.P.; et al. Efficient assembly of nanopore reads via highly accurate and intact error correction. *Nat. Commun.* **2021**, *12*, 60. [CrossRef] [PubMed]
38. Walker, B.J.; Abeel, T.; Shea, T.; Priest, M.; Abouelliel, A.; Sakthikumar, S.; Cuomo, C.A.; Zeng, Q.; Wortman, J.; Young, S.K.; et al. Pilon: An integrated tool for comprehensive microbial variant detection and genome assembly improvement. *PLoS ONE* **2014**, *9*, e112963. [CrossRef] [PubMed]
39. Stanke, M.; Keller, O.; Gunduz, I.; Hayes, A.; Waack, S.; Morgenstern, B. AUGUSTUS: Ab initio prediction of alternative transcripts. *Nucleic Acids Res.* **2006**, *34*, W435–W439. [CrossRef]
40. Ter-Hovhannisyan, V.; Lomsadze, A.; Chernoff, Y.O.; Borodovsky, M. Gene prediction in novel fungal genomes using an ab initio algorithm with unsupervised training. *Genome Res.* **2008**, *18*, 1979–1990. [CrossRef]
41. Haas, B.J.; Salzberg, S.L.; Zhu, W.; Pertea, M.; Allen, J.E.; Orvis, J.; White, O.; Buell, C.R.; Wortman, J.R. Automated eukaryotic gene structure annotation using EVidenceModeler and the program to assemble spliced alignments. *Genome Biol.* **2008**, *9*, R7. [CrossRef] [PubMed]
42. Huerta-Cepas, J.; Szklarczyk, D.; Heller, D.; Hernández-Plaza, A.; Forslund, S.K.; Cook, H.; Mende, D.R.; Letunic, I.; Rattei, T.; Jensen, L.J.; et al. eggNOG 5.0: A hierarchical, functionally and phylogenetically annotated orthology resource based on 5090 organisms and 2502 viruses. *Nucleic Acids Res.* **2019**, *47*, D309–D314. [CrossRef]
43. Finn, R.D.; Coggill, P.; Eberhardt, R.Y.; Eddy, S.R.; Mistry, J.; Mitchell, A.L.; Potter, S.C.; Punta, M.; Qureshi, M.; Sangrador-Vegas, A.; et al. The Pfam protein families database: Towards a more sustainable future. *Nucleic Acids Res.* **2016**, *44*, D279–D285. [CrossRef] [PubMed]
44. Krzywinski, M.; Schein, J.; Birol, I.; Connors, J.; Gascoyne, R.; Horsman, D.; Jones, S.J.; Marra, M.A. Circos: An information aesthetic for comparative genomics. *Genome Res.* **2009**, *19*, 1639–1645. [CrossRef]
45. Jain, C.; Rodriguez-R, L.M.; Phillippy, A.M.; Konstantinidis, K.T.; Aluru, S. High throughput ANI analysis of 90K prokaryotic genomes reveals clear species boundaries. *Nat. Commun.* **2018**, *9*, 5114. [CrossRef] [PubMed]
46. Wang, Y.; Tang, H.; Debarry, J.D.; Tan, X.; Li, J.; Wang, X.; Lee, T.H.; Jin, H.; Marler, B.; Guo, H.; et al. MCScanX: A toolkit for detection and evolutionary analysis of gene synteny and collinearity. *Nucleic Acids Res.* **2012**, *40*, e49. [CrossRef]
47. Emms, D.M.; Kelly, S. OrthoFinder: Phylogenetic orthology inference for comparative genomics. *Genome Biol.* **2019**, *20*, 238. [CrossRef]
48. Price, M.N.; Dehal, P.S.; Arkin, A.P. FastTree: Computing large minimum evolution trees with profiles instead of a distance matrix. *Mol. Biol. Evol.* **2009**, *26*, 1641–1650. [CrossRef]
49. Zhang, H.; Yohe, T.; Huang, L.; Entwistle, S.; Wu, P.; Yang, Z.; Busk, P.K.; Xu, Y.; Yin, Y. dbCAN2: A meta server for automated carbohydrate-active enzyme annotation. *Nucleic Acids Res.* **2018**, *46*, W95–W101. [CrossRef]
50. Almagro Armenteros, J.J.; Tsirigos, K.D.; Sønderby, C.K.; Petersen, T.N.; Winther, O.; Brunak, S.; von Heijne, G.; Nielsen, H. SignalP 5.0 improves signal peptide predictions using deep neural networks. *Nat. Biotechnol.* **2019**, *37*, 420–423. [CrossRef]
51. Chen, Y.; Yu, P.; Luo, J.; Jiang, Y. Secreted protein prediction system combining CJ-SPHMM, TMHMM, and PSORT. *Mamm. Genome* **2003**, *14*, 859–865. [CrossRef]
52. Urban, M.; Cuzick, A.; Seager, J.; Wood, V.; Rutherford, K.; Venkatesh, S.Y.; De Silva, N.; Martinez, M.C.; Pedro, H.; Yates, A.D.; et al. PHI-base: The pathogen-host interactions database. *Nucleic Acids Res.* **2020**, *48*, D613–D620. [CrossRef]
53. Lu, T.; Yao, B.; Zhang, C. DFVF: Database of fungal virulence factors. *Database* **2012**, *2012*, bas032. [CrossRef] [PubMed]
54. Banks, A.M.; Aminuddin, F.; Williams, K.; Batstone, T.; Barker, G.L.A.; Foster, G.D.; Bailey, A.M. Genome sequence of *Lecanicillium fungicola* 150-1, the causal agent of dry bubble disease. *Microbiol. Resour. Ann.* **2019**, *8*, e00340-19. [CrossRef] [PubMed]
55. Kumar, A.; Sharma, V.P.; Kumar, S.; Nath, M. *De novo* genome sequencing of mycoparasite *Mycogone perniciosa* strain MgR1 sheds new light on its biological complexity. *Braz. J. Microbiol.* **2021**, *52*, 1545–1556. [CrossRef]
56. Christinaki, A.C.; Myridakis, A.I.; Kouvelis, V.N. Genomic insights into the evolution and adaptation of secondary metabolite gene clusters in fungicolous species *Cladobotryum mycophilum* ATHUM6906. *G3* **2024**, *14*, jkae006. [CrossRef]
57. Liu, Q.; Xu, Y.; Zhang, X.; Li, K.; Li, X.; Wang, F.; Xu, F.; Dong, C. Infection process and genome assembly provide insights into the pathogenic mechanism of destructive mycoparasite *Calcarisporium cordycipiticola* with host specificity. *J. Fungi* **2021**, *7*, 918. [CrossRef] [PubMed]
58. Yuan, X.; Peng, K.; Li, C.; Zhao, Z.; Zeng, X.; Tian, F.; Li, Y. Complete genomic characterization and identification of *Saccharomycopsisphalluae* sp. nov., a novel pathogen causes yellow rot disease on *Phallus rubrovolvatus*. *J. Fungi* **2021**, *7*, 707. [CrossRef]
59. Manni, M.; Berkeley, M.R.; Seppey, M.; Simão, F.A.; Zdobnov, E.M. BUSCO update: Novel and streamlined workflows along with broader and deeper phylogenetic coverage for scoring of eukaryotic, prokaryotic, and viral genomes. *Mol. Biol. Evol.* **2021**, *38*, 4647–4654. [CrossRef]
60. Liu, L.; Tang, M.C.; Tang, Y. Fungal highly reducing polyketide synthases biosynthesize salicylaldehydes that are precursors to epoxycyclohexenol natural products. *J. Am. Chem. Soc.* **2019**, *141*, 19538–19541. [CrossRef]

61. Utermark, J.; Karlovsky, P. Role of zearalenone lactonase in protection of *Gliocladium roseum* from fungitoxic effects of the mycotoxin zearalenone. *Appl. Environ. Microbiol.* **2007**, *73*, 637–642. [CrossRef] [PubMed]
62. Langner, T.; Göhre, V. Fungal chitinases: Function, regulation, and potential roles in plant/pathogen interactions. *Curr. Genet.* **2016**, *62*, 243–254. [CrossRef] [PubMed]
63. Inbar, J.; Chet, I. The role of recognition in the induction of specific chitinases during mycoparasitism by *Trichoderma harzianum*. *Microbiology* **1995**, *141*, 2823–2829. [CrossRef] [PubMed]
64. Yang, Y.; Sossah, F.L.; Li, Z.; Hyde, K.D.; Li, D.; Xiao, S.; Fu, Y.; Yuan, X.; Li, Y. Genome-wide identification and analysis of chitinase GH18 gene family in *Mycogone perniciosa*. *Front. Microbiol.* **2021**, *11*, 596719. [CrossRef] [PubMed]
65. Calonje, M.; Mendoza, C.G.; Cabo, A.P.; Bernardo, D.; Novaes-Ledieu, M. Interaction between the mycoparasite *Verticillium fungicola* and the vegetative mycelial phase of *Agaricus bisporus*. *Mycol. Res.* **2000**, *104*, 988–992. [CrossRef]
66. Abubaker, K.S.; Sjaarda, C.; Castle, A.J. Regulation of three genes encoding cell-wall-degrading enzymes of *Trichoderma aggressivum* during interaction with *Agaricus bisporus*. *Can. J. Microbiol.* **2013**, *59*, 417–424. [CrossRef]
67. Mills, P.R.; Fermor, T.; Muthumeenakshi, S.; Lincoln, S. Cell wall degrading enzymes produced by *Verticillium* spp. and their relationship to infection in *Agaricus bisporus*. In Proceedings of the 15th International Congress on the Science and Cultivation of Edible Fungi, Maastricht, The Netherlands, 15–19 May 2000; Volume 15, pp. 601–605.
68. Xie, Y.R.; Raruang, Y.; Chen, Z.Y.; Brown, R.L.; Cleveland, T.E. ZmGns, a maize class I β-1,3-glucanase, is induced by biotic stresses and possesses strong antimicrobial activity. *J. Integr. Plant Biol.* **2015**, *57*, 271–283. [CrossRef]
69. Sjaarda, C.P.; Abubaker, K.S.; Castle, A.J. Induction of *lcc2* expression and activity by *Agaricus bisporus* provides defence against *Trichoderma aggressivum* toxic extracts. *Microb. Biotechnol.* **2015**, *8*, 918–929. [CrossRef]
70. Lu, M.; Wen, T.; Guo, M.; Li, Q.; Peng, X.; Zhang, Y.; Lu, Z.; Wang, J.; Xu, Y.; Zhang, C. Regulation of intracellular reactive oxygen species levels after the development of *Phallus rubrovolvatus* rot disease due to *Trichoderma koningii* mycoparasitism. *J. Fungi* **2023**, *9*, 525. [CrossRef]
71. Brown, D.W. The KP4 killer protein gene family. *Curr. Genet.* **2011**, *57*, 51–62. [CrossRef]

foods

MDPI

Article

The Effect of Cold Plasma Treatment on the Storage Stability of Mushrooms (*Agaricus bisporus*)

Yalong Guo [1], Shuqiong Xia [1], Chong Shi [2], Ning Ma [3], Fei Pei [3], Wenjian Yang [3], Qiuhui Hu [3], Benard Muinde Kimatu [4] and Donglu Fang [2,*]

1 College of Light Industry and Food Engineering, Nanjing Forestry University, Nanjing 210037, China; guoyalong@njfu.edu.cn (Y.G.); x16211357@163.com (S.X.)
2 State Key Laboratory of Tree Genetics and Breeding, Co-Innovation Center for Sustainable Forestry in Southern China, College of Forestry and Grassland, Nanjing Forestry University, Nanjing 210037, China; shichong0639@163.com
3 Jiangsu Province Engineering Research Center of Edible Fungus Preservation and Intensive Processing, College of Food Science and Engineering, Nanjing University of Finance and Economics, Nanjing 210023, China; maning@nufe.edu.cn (N.M.); feipei87@163.com (F.P.); lingwentt@163.com (W.Y.); qiuhuihu@nufe.edu.cn (Q.H.)
4 Department of Dairy and Food Science and Technology, Egerton University, Egerton 20115, Kenya; muinde.bk@gmail.com
* Correspondence: fangdonglu@njfu.edu.cn

Abstract: Postharvest *Agaricus bisporus* is susceptible to browning, water loss, and microbial infection. In order to extend its shelf life, cold plasma technology was used to treat and evaluate *A. bisporus*. Firstly, according to the results of a single factor test and response surface analysis, the optimal conditions for cold plasma treatment were determined as a voltage of 95 kV, a frequency of 130 Hz, and a processing time of 10 min. Secondly, storage experiments were carried out using the optimized cold plasma treatment. The results showed that the cold plasma treatment in the packaging significantly reduced the total viable count in *A. bisporus* by approximately 16.5%, maintained a browning degree at 26.9% lower than that of the control group, and a hardness at 25.6% higher than that of the control group. In addition, the cold plasma treatment also helped to preserve the vitamin C and total protein content of *A. bisporus*. In conclusion, cold plasma treatment showed great potential in enhancing the postharvest quality of fresh *A. bisporus*.

Keywords: *Agaricus bisporus*; cold plasma; response surface; optimization; postharvest quality

1. Introduction

Agaricus bisporus, commonly known as the white mushroom [1], is a fungus that has gained global attention due to its unique sensory and nutritional values [2]. However, *A. bisporus* has a short shelf life under natural postharvest conditions owing to its high water content, lack of protective surface cuticle, excessive respiration and transpiration, and microbial infection [3]. These factors contribute to the browning, water loss, softening, nutrient loss, and spoilage of postharvest *A. bisporus*. In recent years, various preservation methods have been explored to address these challenges, including improved packaging techniques [4–6], light treatment [7], and irradiation treatment [8]. However, these methods have limitations such as the complexity in preparing active packaging, the high cost of raw materials, and difficulties in mass production. Alternatively, single low-temperature preservation methods require significant investment and have low efficiency. Therefore, researchers have increasingly focused on developing new preservation technologies, such as cold plasma combined with a refrigerated packing treatment, which have shown promising results in the preservation and processing of various foods [9,10].

Cold plasma is a novel non-contact cold sterilization technology that exhibits broad-spectrum sterilization properties [11]. Its bactericidal efficacy has been demonstrated in various food products including *Flammulina velutipes* [12], *Agaricus bisporus* [13], winter jujubes [14], apricot [15], and fresh wolfberries [16]. Plasma, a highly active substance rich in reactive oxygen and nitrogen [17], can permeate the microbial cell membrane, leading to DNA damage and a reduction in the activity of cell-degrading enzymes [18]. Moreover, these active substances also impact on the quality and properties of biomolecules [17]. For instance, cold plasma has the ability to depolymerize amylose [19], disrupt peptide bonds in proteins [20], and promote lipid oxidation [21].

Cold plasma can be generated using various methods. Among these methods, dielectric barrier discharge (DBD) is particularly well-suited for food industrial applications [22]. DBD offers several advantages, including the ability to produce large-area and uniform plasma, short processing times, high energy efficiency, high antimicrobial efficiency, and minimal impact on food quality and the environment [23]. However, due to insufficient research on cold plasma technology and the lack of theoretical foundations, its application in the field of food preservation is not widespread, especially for mushroom preservation.

This study aimed to improve the storage period of *A. bisporus* by optimizing the treatment conditions of cold plasma and investigating its effects on the postharvest physiological and nutritional quality of the mushroom. The effects of cold plasma voltage, frequency, and processing time on the surface microorganisms and color were examined during the storage of *A. bisporus* using the Box-Behnken response surface method and a quadratic polynomial regression model. Additionally, the postharvest physiological characteristics and nutritional changes in *A. bisporus* after cold plasma treatment were evaluated. These findings would provide valuable insights into the application of cold plasma treatment on edible mushrooms.

2. Materials and Methods

2.1. Sample Source and Pre-Treatment

A. bisporus samples were provided by Jiangsu zhongyouxinghe mushroom technology Co., Ltd. (Lianyungang, China) and promptly sent to the laboratory within 2 h of harvest. To ensure the freshness of the edible fungi, only unopened mushrooms without any mechanical damage were chosen. The selected mushrooms had a cap diameter of approximately 40 mm and weighed between 18–25 g. Prior to testing, the *A. bisporus* samples were pre-cooled at 4 °C and 90% relative humidity for 24 h. The selected *A. bisporus* samples were randomly packed, with each box containing approximately 120 g. The boxes were placed in food-grade polypropylene PP boxes measuring 18 cm × 13 cm × 6 cm.

2.2. Equipment and Reagents

This study used cold plasma equipment from Nanjing Suman (CPS-1, Nanjing Suman Plasma Technology Co., Ltd., Nanjing, China). *A. bisporus* was placed on a glass plate (dielectric barrier layer) among the electrodes. The size of electrodes was 17 cm × 13 cm.

Folinol and gallic acid (CAS: 149-91-7) were purchased from Shanghai yuanye Bio-Technology Co., Ltd. (Shanghai, China). Guaiacol (CAS: 90-05-1), phthalic acid, diphenols, ergosterol (CAS: 57-87-4), and vitamin D2 (CAS: 50-14-6) were purchased from Shanghai Macklin Biochemical Technology Co., Ltd. (Shanghai, China). Methanol and other reagents were purchased from Sinopharm chemical reagent Co., Ltd. (Shanghai, China). Except for the liquid chromatography-specific reagents, these reagents were of analytical grade and used without further purification.

2.3. Cold Plasma Process Optimization

2.3.1. Single Factor Experiment

As shown in Figure S1A,B, *A. bisporus* was sealed in a polyethylene film bag with a thickness of 0.07 mm. It was then promptly placed on the electrode of a cold plasma device for further processing. During the experiment, some fixed parameters were set as follows: the number of discharges was once, the distance between the plates was 60 mm, the upper limit of the current was 1.5 A, and the temperature was room temperature: 25 ± 5 °C.

When the single factor experiment was carried out, the other factors were fixed as follows: treatment voltage 125 kV, treatment frequency 145 Hz, processing time 7 min.

This study was based on the cold plasma treatment voltage (50, 65, 80, 95, 110, 125, 140, 155 kV), treatment frequency (70, 95, 120, 145, 170, 195 Hz) and processing time (1, 4, 7, 10, 13, 16 min) as the optimized conditions, and the samples were processed in the sealed box. Each treatment was repeated three times. *A. bisporus* without cold plasma treatment was used as a control. After treatment, *A. bisporus* was placed in a constant temperature and humidity box at 4 °C and 90% relative humidity and stored for 24 h. Please refer to the methods described in Sections 2.5.2 and 2.5.3 for the determination of sterilization rate and color difference.

For the detailed results in the single factor experiment, please refer to the Supplementary Information and Figures S2–S4.

2.3.2. Box-Behnken Experiment for Optimization

The optimization of the cold plasma treatment conditions for *A. bisporus* was conducted using the Box-Behnken design. The variables considered were voltage, frequency, and time, while the responses measured were the surface L* of *A. bisporus* and the sterilization rate, denoted as Y. Prior to the experiment, the range of each variable was determined through a single factor experiment, and other parameters of the equipment were the same as shown in Section 2.3.1. A total of seventeen experiments were performed based on the Box-Behnken design matrix, as detailed in Table S1. In each experimental group, *A. bisporus* samples were treated with cold plasma using the specified voltage, frequency, and time settings. Subsequently, the surface L* and sterilization rate were measured after the treatment.

2.4. Cold Plasma Treatment and Storage of *A. bisporus*

As shown in Section 2.1, the packaged *A. bisporus* were randomly divided into four groups (Table 1 and Figure S1). They were the following four groups: cold plasma treatment combined with PE sealed packaging (CP+PE), PE sealed packaging (PE), direct cold plasma treatment (CP), and control group (Control). The packaging group used 0.07 mm thick polyethylene film bags for the sealed packaging. Samples were stored at 4 °C and 90% relative humidity for 12 days, and the physical and chemical indicators were measured and evaluated on days 0, 3, 6, 9 and 12 [24].

Table 1. Process conditions for the treatment of *Agaricus bisporus*.

Treatment	Packaging	Plasma Treatment
CP+PE	Yes	Yes
PE	Yes	No
CP	No	Yes
Control	No	No

2.5. Preservation Quality Investigation

2.5.1. Hardness, Weight Loss Rate, and Moisture Content

The hardness was measured using a fruit firmness tester with a 3.5 mm flat probe (GY-4; Jinkelida Instrument Co., Ltd., Beijing, China) and expressed as N [25].

Weight loss was measured by weighing samples before and after storage. The results were expressed as the percentage of weight lost relative to the initial weight.

The moisture content was determined by a direct drying method. First, the weighing bottle was dried to a constant weight and the weight (W_1) was recorded. Secondly, approximately 5 g of chopped mushroom samples were added, and the total weight (W_2) of the sample and the weighing bottle was recorded in detail. Finally, the sample was dried to a constant weight and the total weight (W_3) after drying was recorded. The formula for moisture content is as follows:

$$X(\%) = \frac{W_2 - W_3}{W_2 - W_1} \times 100\%$$

2.5.2. Microbial Colony Assay of Mushroom Epidermis

The total number of colonies was measured according to a previously described method with minor modifications [8]. *A. bisporus* (25 g) from each group were placed into a sterile conical flask, to which sterile saline (225 mL) was added. Then, the solution was vortexed for 2 min. The solution (100 μL) was then deposited onto plate count agar plates and evenly spread and incubated for 48 h at 37 °C for the aerobic colony count. Each sample had three replicates. The results were expressed as the number of colonies per gram of fresh weight (lg CFU g^{-1}), and lg was logarithmic to the base of 10.

$$Y(\%) = (1 - \frac{M_1}{M_2}) \times 100\%$$

Y(%) was the sterilization rate.
M_1 was the number of colonies after low temperature plasma treatment.
M_2 was the number of untreated colonies.

2.5.3. Mushroom Appearance, Color, Peroxidase, and Polyphenol Oxidase

The mushrooms were photographed at the beginning and at the end of the storage period [26]. The color was recorded as L*, a*, and b* using a 3NH colorimeter (3NH SR-66; Shenzhen 3NH Technology Co., Ltd., Shenzhen, China). The browning index (BI) was determined according to the following formula:

$$BI = [100 \times (X - 0.31)] \quad (1)$$

$$X = \frac{a^* - 1.75L^*}{5.546L^* + a^* - 3.012b^*} \quad (2)$$

To determine enzyme activities, the mushrooms (5 g) were added to 5 mL of extraction buffer containing 1 mmol L^{-1} IPEG, 4% PVPP, and 1% Triton X-100. The extract was centrifuged at 4 °C, 12,000× *g* for 30 min.

Peroxidase (POD) activity was measured according to the method of Yan with some modifications [5]. Firstly, 0.5 mL of the enzyme extract was mixed with 3.0 mL of 25 mmol L^{-1} guaiacol, and afterwards 200 μL of 0.5 mol L^{-1} H_2O_2 was added and quickly mixed. The absorbance of the reaction mixture was then measured six times at 470 nm with 1 min intervals. One unit (U) of POD activity was defined as a change of 0.01 at 470 nm per minute. POD activity was expressed as U g^{-1} fresh weight.

Polyphenol oxidase (PPO) was measured according to the method of Yan [5] with some modifications. Firstly, 4.0 mL of 50 mmol L^{-1} sodium acetate buffer solution and 1.0 mL of 50 mmol L^{-1} catechol were mixed, and then 100 μL of the enzyme extract was added and mixed well. The absorbance of the reaction mixture was finally measured six times at 420 nm with 1 min intervals. One unit (U) of PPO activity was defined as a change of 0.01 at 420 nm per minute. PPO activity was expressed as U g^{-1} fresh weight.

2.5.4. Vitamin C, Total Protein Content, Total Phenolic Content, and Total Soluble Solids Content

The amount of vitamin C and total protein were determined using kits from Nanjing Jiancheng Biotechnology Co., Ltd. (Nanjing, China).

The total phenolic content (TPC) of the samples was detected by the Folin-Ciocalteu method [24], and gallic acid was used to construct a standard curve. Absorbance of the supernatant was measured at 760 nm and the mass fraction of total phenol in the samples (mg g^{-1}) was calculated.

The determination of total soluble solids (TSS) was based on Ni's method [27] with some minor modifications. The mushroom samples (10 g) were added to 15 mL of deionized water (pH 7), then ground in a mortar and filtered with four layers of gauze to remove the filter residue. Subsequently, TSS (%) was measured via a refractometer (Atago PAL-1; Atago Co. Ltd., Tokyo, Japan) using the supernatant.

2.5.5. Determination of Vitamin D and Ergosterol

A direct extraction method was employed to extract vitamin D and ergosterol from the mushrooms (Nzekoue et al., 2022) [28]. Mushroom powder (0.2 g) was subjected to ultrasonic extraction in ethanol (5 mL) for 30 min. The ultrasonic extraction conditions included a power of 240 W, frequency of 40 kHz, and temperature of 40 °C. Following ultrasonication, the extract was filtered using a 0.45 μm organic microporous membrane and transferred to a liquid phase test bottle for subsequent HPLC analysis.

The HPLC system (Waters alliance e2695 Series, Milford, MA, USA) was equipped with a quaternary pump, on-line degasser, thermostatic autosampler, and diode array detector (DAD). The chromatographic separation was performed on a Gemini C18 analytical column (250 × 3.0 mm, 5 μm) preceded by a security guard column C18 (4 × 3 mm, 5 μm). The mobile phase was A phase composed of water and B phase composed of methanol, and the flow rate was 0.8 mL min^{-1}. The gradient elution was started with 80% B phase for 5 min, increased to 100% in 6 min, maintained for 13 min, and returned to the initial conditions within 1 min. The injection volume was 20 μL and the column temperature was 40 °C. Vitamin D_2 and ergosterol were detected at 265 nm and 280 nm, respectively, and used as a calibration curve for quantitation.

2.6. *Statistical Analyses*

All indicators were measured in triplicate. The data were expressed as mean ± SD, and the experimental data were organized and plotted using Origin Pro 2024b (OriginLab, Northampton, MA, USA). Analysis of variance (ANOVA) was performed using SPSS 26 software (IBM, Armonk, NY, USA). To determine the statistical differences, comparisons of the means between the controls and the treatments were performed using Duncans test at a significance level of $p < 0.05$.

3. Results and Discussion

3.1. *Optimization of Cold Plasma Treatment for A. bisporus*

High-intensity treatments have been found to potentially disrupt the surface structure of agricultural products and oxidize antioxidant substances [16]. In this study, we aimed to optimize the treatment conditions using a response surface methodology. It is important to consider that voltage can cause air discharge effects, frequency may impact energy input and product structure, and processing time determines the total treatment energy during cold plasma processing [29]. Therefore, we selected voltage, frequency, and time as the main factors for the response surface analysis. Our goal was to find suitable treatment conditions that effectively kill pathogenic microorganisms while minimizing browning and nutritional damage in *A. bisporus*. To prepare for the response surface analysis, we conducted a preliminary single factor experiment to assess the influence of voltage, frequency, and time on the surface L* and sterilization rate (Y) of *A. bisporus* (Figures S1–S3). Table S1 presented the design and results of the Box-Behnken experiments. Using DesignExpert 8.0 software, we performed quadratic polynomial regression fitting of the sterilization rate data in Table S1 to obtain a multiple regression equation describing the relationship between the sterilization rate and the independent variables.

In Table 2, the L* regression model is not significant, and the sterilization rate Y regression model is extremely significant ($p < 0.01$). The lack of fit term of the sterilization rate Y regression model is not significant ($p = 0.6966 > 0.05$), and the coefficient of determination ($R^2 = 0.9631$) and the corrected coefficient of determination ($R^2_{Adj} = 0.9156$) of the equation is high, indicating that the fitting degree of the model is high and the experimental error is small.

Improper handling conditions can have a significant impact on the freshness of mushrooms by promoting biological processes like respiration, ripening, and senescence, and enabling pathogenic microorganisms. To analyze the impact of different experimental factors on the response variables, we considered the processing time, treatment frequency, and voltage. The F values indicated the importance of each factor, with processing time having the least influence, followed by treatment frequency, and voltage being the most influential. In Figure 1A–C, we present three-dimensional response surface diagrams to visualize the interaction of these factors with the response variables. The shape of the response surface indicates the extent of interaction, with a greater curvature suggesting a more significant interaction.

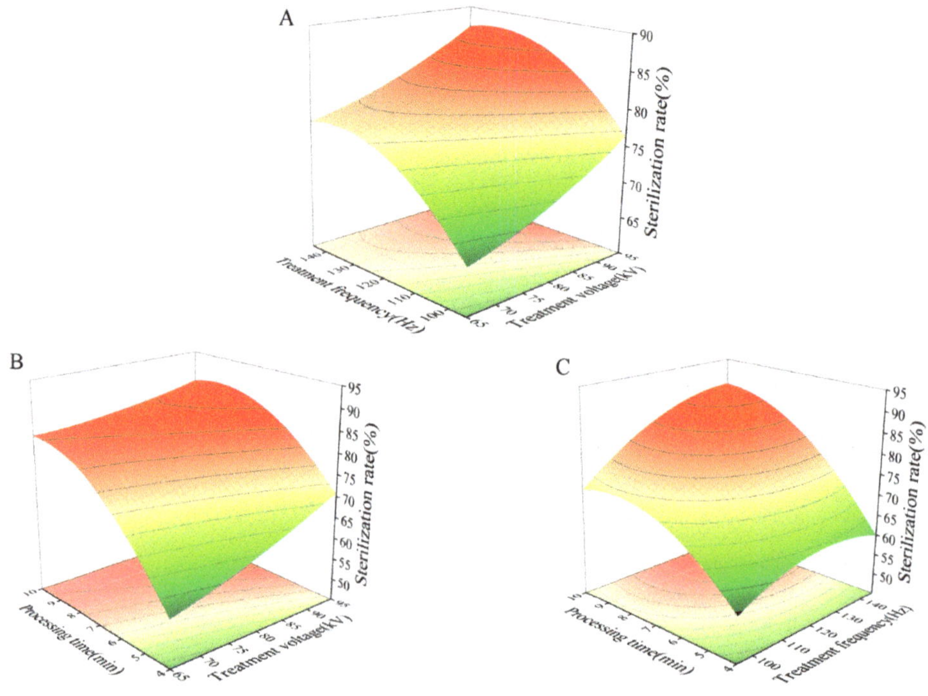

Figure 1. Response surface diagrams showing the interactions of different factors on the sterilization rate. (**A**) Effects of treatment voltage and frequency on the sterilization rate of *A. bisporus*. (**B**) Effects of treatment voltage and time on the sterilization rate of *A. bisporus*. (**C**) Effects of treatment time and frequency on the sterilization rate of *A. bisporus*.

The predicted optimal treatment conditions were determined based on the highest bactericidal rate Y. The predicted optimum treatment conditions were 95.154 kV, 130.41 Hz, and 9.612 min, adjusted to 95 kV, 130 Hz, and 10 min for operational purposes. These conditions were verified, and the actual Y was 89.05% with an average relative error of ±1.744%, indicating reliable model optimization results. Many studies have reported the effects of treatment parameters on the preservation quality of edible fungi. Excessive application of parameters usually caused higher oxidative damage to mushrooms and reduced the edible quality of mushrooms [19].

Table 2. Analysis of variance of regression equations for two response variables.

Source	Y			L*		
	F-Value	p-Value		F-Value	p-Value	
Model	20.29	0.0003	significant	1.39	0.3079	not significant
A-A	20.76	0.0026		5.19	0.0459	
B-B	25.24	0.0015		0.0004	0.9851	
C-C	99.1	<0.0001		0.0003	0.9857	
AB	0.0028	0.9589		1.26	0.2887	
AC	2.19	0.1826		0.8428	0.3802	
BC	5.03	0.0597		1.04	0.3316	
A^2	0.5301	0.4902				
B^2	7.97	0.0257				
C^2	20.74	0.0026				
Lack of Fit	0.51	0.6966	not significant	0.0411	0.9992	not significant
R^2	0.9631			0.4544		
Adjusted R^2	0.9156			0.1271		
Predicted R^2	0.7948			0.2984		

3.2. *Postharvest Preservation of A. bisporus*

3.2.1. Hardness, Weight Loss Rate, Moisture Content, and Total Number of Bacterial Colonies

According to Figure 2A, the polyethylene packaging group gradually softens throughout the postharvest storage process, while the unpackaged group remain unchanged in hardness. This could be due to the severe water loss and shrinkage of the mushroom surface in the unpackaged group, resulting in the formation of a dense waterproof layer on the mushroom surface. On the ninth day of storage, the hardness of the CP+PE group remained at 13.79 N, which was significantly higher ($p < 0.05$) than that of the PE group. This indicated that the cold plasma treatment had the potential to delay the softening of the mushrooms [30]. This delay in softening might be due to the inactivation of bacteria by active plasma species, inhibition of enzyme activity, and a reduction in the respiratory rate [13].

The rate of water transpiration and the respiration rate in mushrooms during storage are two common issues leading to weight loss [31]. In Figure 2B, the weight loss in unpackaged mushrooms significantly increases ($p < 0.05$) with an extended storage time. After three days of storage, the weight loss in samples from the CP and control groups exceeded the permitted limit, with values of 9.68% and 8.11%, respectively. Mushrooms were considered to have lost their value when their water loss reached 5% [32]. However, we observed condensation on the surface and inside of the mushrooms in the packaging group, which might be attributed to the low water vapor permeability of the low-density polyethylene film. This condensation phenomenon could potentially contribute to the deterioration of the mushrooms [33]. The moisture content of *A. bisporus* is an important indicator of freshness. Figure 2C demonstrates that the moisture content of unpackaged mushrooms decreases after 12 days of storage, consistent with the results of the weight loss rate.

In this study, the total number of bacteria was used to monitor the bactericidal effect of cold plasma on *A. bisporus*. The results in Figure 2D demonstrate an upward trend in the number of microorganisms during the storage period. Towards the later stage of storage, the CP+PE group exhibited significantly lower microorganism counts compared to the PE group ($p < 0.05$). These findings indicate that the cold plasma treatment effectively reduces microorganism levels on the surface of *A. bisporus*, thereby enhancing mushroom safety. In addition, similar results were also exhibited in the research of Subrahmanyam, K, which showed that the number of bacteria and fungi in *A. bisporus* decreased significantly after cold plasma treatment ($p < 0.05$) [13]. Pourbagher, R used surface dielectric barrier discharge plasma to inactivate *P. tolaasii Pt18* bacteria inoculated into *A. bisporus*. H_2O_2 + air gas had the greatest effect on reducing bacteria, and the number of colonies decreased to 4.23 lg CFU g^{-1} on the 21st day of storage [34]. It was worth noting that the bactericidal performance of cold plasma was attributed to various active substances which caused lipid peroxidation, enzyme inactivation, and DNA degradation, eventually leading to microbial inactivation [35].

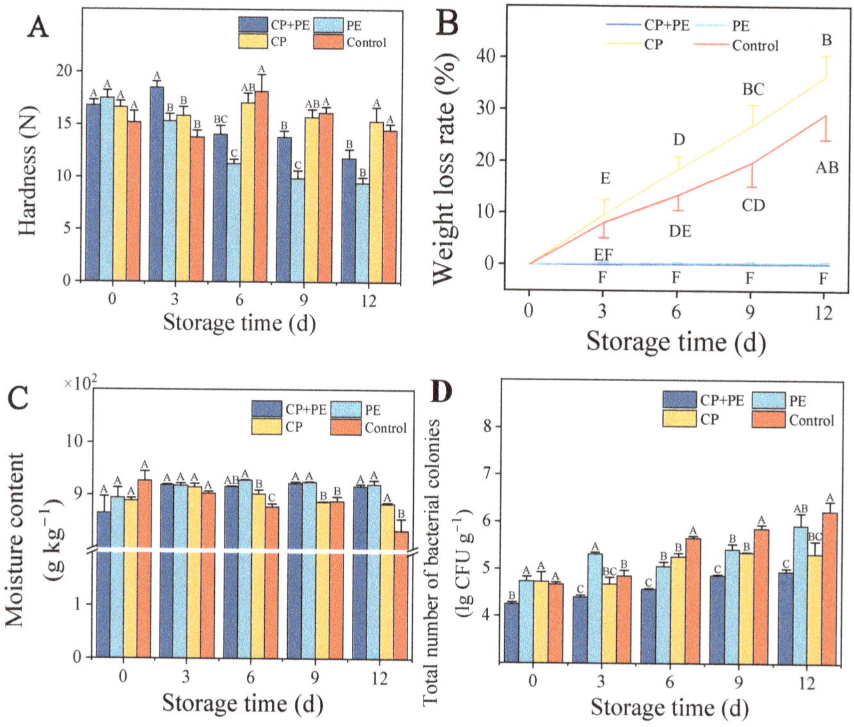

Figure 2. Effects of different treatments on hardness (**A**), weight loss rate (**B**), moisture content (**C**), and total number of colonies (**D**) of *Agaricus bisporus*. Distinct letters demonstrate a significant difference between different treatments ($p < 0.05$).

3.2.2. Appearance, Color, Peroxidase, and Polyphenol Oxidase

Changes in the mushroom surface color play a crucial role in determining the shelf life of *A. bisporus*, and are a significant factor in influencing consumer acceptance [27,36]. A decrease in the L* value indicates a darker colored *A. bisporus*. Figure 3A highlights an important pattern where the L* value of *A. bisporus* progressively decreases with a longer storage time. The slowest color change is observed in *A. bisporus* treated with cold plasma and packed in polyethylene. These color change indices are further supported by the mushroom appearance images at different storage periods (Figure 4). After nine days of cold storage, the PE group samples exhibit black spots and noticeable browning, whereas the CP+PE treated group maintain a better color (Figure 3B). Cold plasma treatment in packaging effectively inhibited browning. Zhao's research indicates that plasma activated water (PAW) treatment could effectively suppress browning and retain the desired visual appearance of button mushrooms [37]. The increase in browning values and the decrease in lightness values of mushrooms could be due to microbial growth, water loss, and enzymatic browning [13,35].

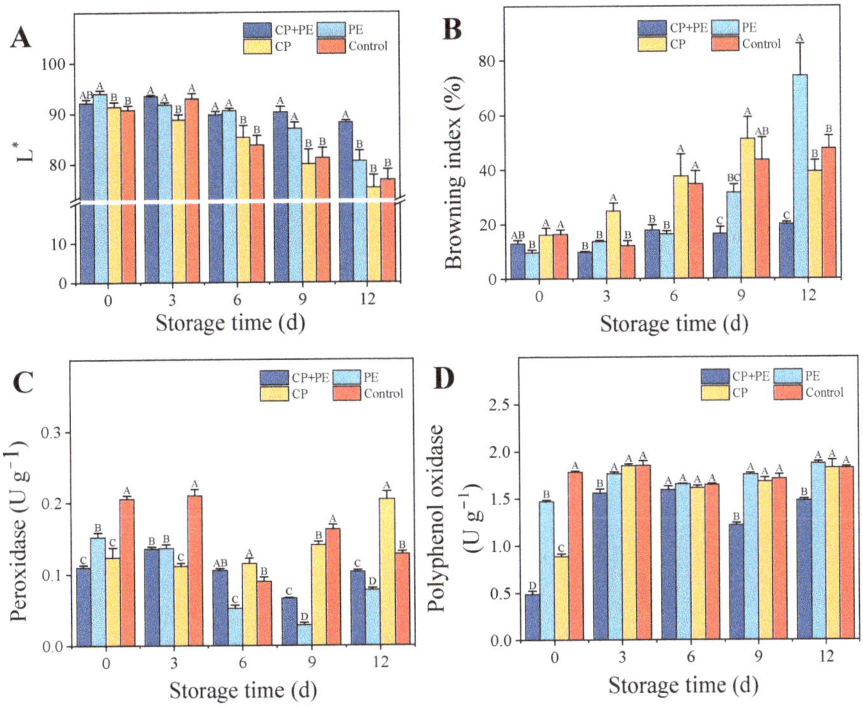

Figure 3. Effects of different treatments on L* (**A**), browning index (**B**), polyphenol oxidase (**C**), and peroxidase (**D**) of *Agaricus bisporus*. Distinct letters demonstrate a significant difference between different treatments ($p < 0.05$).

Enzymatic browning is the main type of *A. bisporus* browning, which is generally considered to be brown spots formed due to the oxidation of phenols by PPO and POD [38]. Cold plasma treatment has been shown to inactivate enzymes [39]. Following the initial treatment on day 0, the activities of peroxidase (POD) and polyphenol oxidase (PPO) in the cold plasma treatment group are significantly decreased (Figure 3C,D). This decrease in enzyme activity could be attributed to a reduction in the α-helix structure and an increase in the β-sheet structure within the enzyme proteins' secondary structure [40]. However, during the later storage period, this trend is reversed. The POD activity in the CP+PE group is significantly higher than that of PE group ($p < 0.05$), whereas the PPO activity is significantly lower ($p < 0.05$). Roghayeh Pourbagher's research indicated that, under dual dielectric barrier discharge plasma treatment, the POD activity in *A. bisporus* decreased, and this decreasing trend aligned with the inactivation trend of PPO [41]. Both enzymes decreased with an increase in processing time. This finding is consistent with the trend observed in *A. bisporus* that we processed on day 0. However, as the storage time increases, the POD activity in the CP+PE treatment group is significantly higher than that in the PE group. This discrepancy may be related to the equipment parameters used, specifically their design, with a fixed frequency of up to 6200 Hz. Previous studies indicated that low-power and short-term cold plasma treatment did not decrease POD activity, but rather enhanced it [42]. This finding was confirmed by the preservation of acai pulp and blueberries [43,44]. However, the exact mechanism underlying the activity of the POD enzyme remained unclear [45].

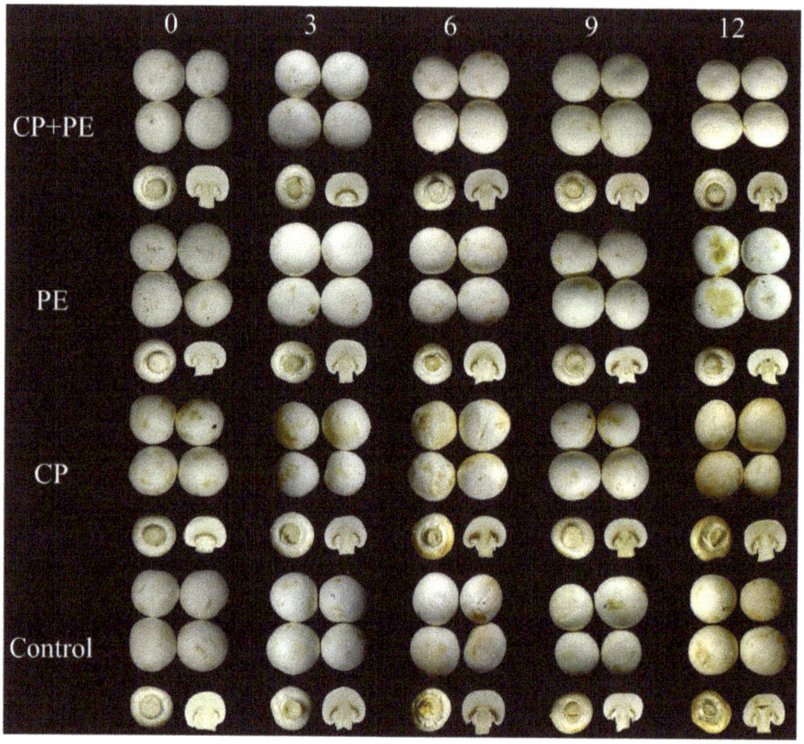

Figure 4. Effects of different treatments on the appearance of *Agaricus bisporus*.

3.2.3. Vitamin C, Protein Content, Total Phenolic Content, and Total Soluble Solids

Vitamin C, an essential nutrient, plays a crucial role in protecting DNA from the damage caused by free radicals [46]. The vitamin C content of the CP+PE group decreases during storage (Figure 5A). This decline could be attributed to the production of numerous active substances by the cold plasma treatment, prompting *A. bisporus* to generate more vitamin C as a defense mechanism against oxidative damage. Over time, the active substances in the packaging gradually decomposed, leading to a stabilization of the vitamin C content in *A. bisporus*. On the other hand, the CP and control groups, being exposed to the external environment and environmental microorganisms, consistently maintained high levels of vitamin C. The vitamin C content in the PE group remained relatively stable throughout. The treatment of cashew apple juice with cold plasma resulted in an increase in the vitamin C content [47], which was considered beneficial. This increase could be attributed to the activation of the enzyme dehydroascorbate reductase, which converts dehydroascorbic acid back to ascorbic acid through the ascorbate–glutathione cycle. Similarly, in a study on bananas, cold plasma treatment caused an increase in the vitamin B_6 content, as the treatment disrupted the cell wall and released the contents [48]. In Ding's study, cold plasma treatment could effectively maintain the vitamin C content of *Flammulina velutipes* during storage, which was significantly higher than that of the control group ($p < 0.05$) [12]. However, Kadavakollu Subrahmanyam's study pointed out that during the storage period, there were no significant differences in the vitamin C content between different treatment groups ($p > 0.05$) [13]. The difference in this result may be related to the strength of the treatment parameters.

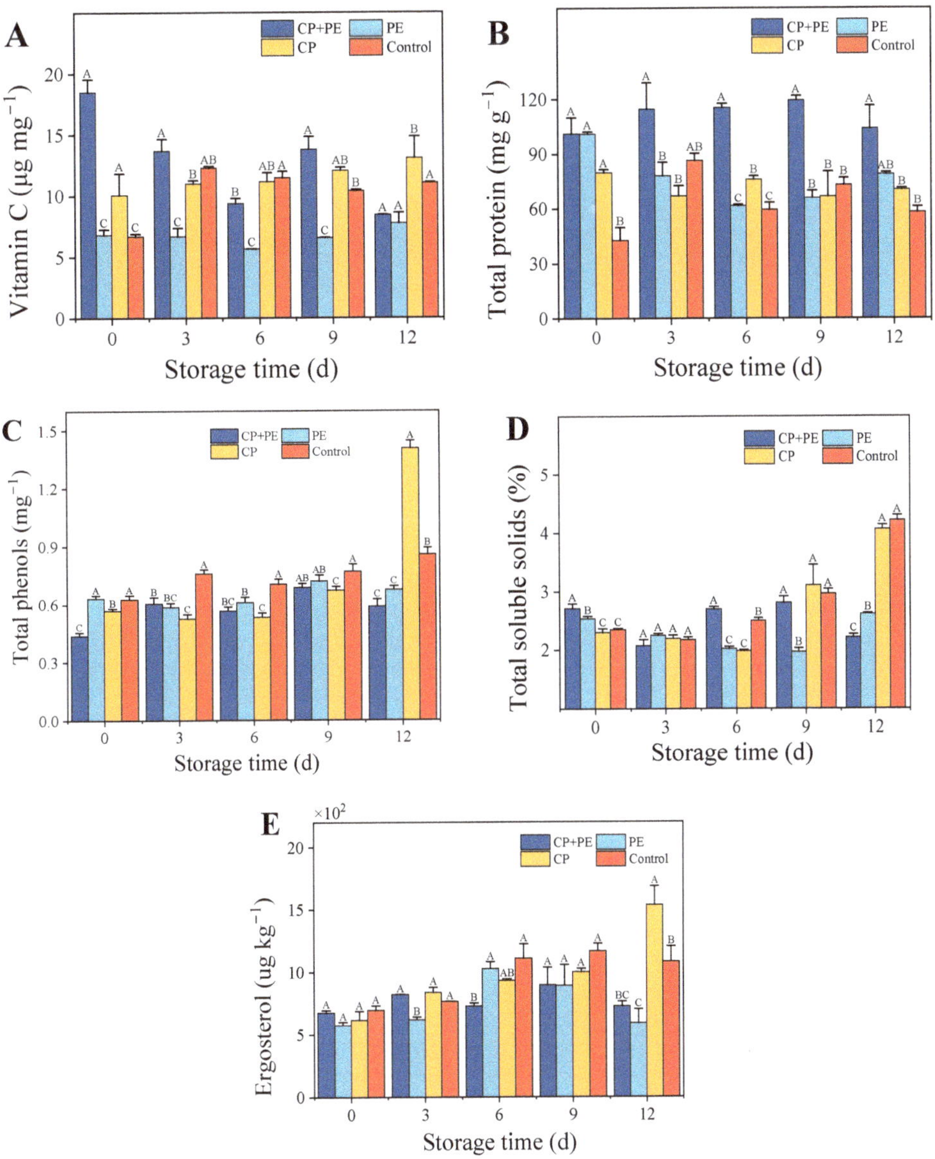

Figure 5. Effects of different treatments on vitamin C (**A**), total protein content (**B**), total phenols (**C**), total soluble solids (**D**), and ergosterol (**E**) of *Agaricus bisporus*. Distinct letters demonstrate a significant difference between different treatments ($p < 0.05$).

Protein content plays a significant role in assessing the nutritional value of mushrooms. In fact, mushroom protein is considered to be of a higher nutritional quality compared to plant protein [49]. The proteins present in mushrooms are involved in various processes such as growth, maturation, senescence, disease resistance, and stress resistance. Figure 5B demonstrates that the soluble protein content in the CP+PE group initially increased and then stabilized, maintaining a higher content throughout the storage period. However, the CP and control groups experienced severe browning and water loss, leading to protein decomposition. The soluble protein content remained relatively low throughout the storage period. In the case of the PE group, the presence of microorganisms resulted in the consumption of protein as a nutrient. It is worth noting that the impact of cold plasma treatment on the structure and function of proteins is well-documented. However, there is a scarcity of literature comparing and explaining the effects of cold plasma on proteins in edible fungi substrates. In this study, cold plasma treatment was found to increase the protein content, which aligned with previous studies on soybean sprouts and black gram seeds [50,51]. This increase could be attributed to the induction of free radicals and ozone oxidation, which promote chemical reactions between amino acid side chains and protein backbones, ultimately resulting in the polymerization of small-molecule proteins [52].

Total phenols are considered beneficial antioxidants due to their ability to scavenge harmful active oxygen species [53]. Figure 5C shows the change in total phenolic content (TPC) during the entire storage period. The CP+PE group exhibits the lowest TPC on day 0. On the ninth day of storage, the PE group reaches its peak TPC. The CP group reaches its highest TPC on the 12th day. Cold plasma treatment could decrease the total phenol content. This decrease is attributed to the ability of phenolic compounds to scavenge free radicals and the reaction of reactive oxygen species with the plasma [43]. Higher treatment intensity leads to the production of ozone and singlet oxygen, which could cleave the central heterocyclic ring in the polyphenol skeleton, leading to a decrease in TPC.

TSS represents the sugar content in the harvested mushrooms, which typically decreases during storage [54]. Limiting the use of soluble solids can help prolong the shelf life of the product. The findings presented in Figure 5D demonstrate noticeable differences in the percentage of soluble solids between packaged and unpackaged mushrooms, which is consistent with the observed changes in the vitamin C content. We obtained similar results in the study of *Flammulina velutipes* [12]. In the packaging group, the TSS content initially decreased and then increased. Before the ninth day of storage, the CP+PE group had a higher TSS content than the PE group, possibly due to a higher microbial content in the PE group, leading to TTS decomposition. On the other hand, the TSS content in the unpackaged group showed a consistent increase throughout the storage period, which might serve as an indicator of mushroom maturity [33].

3.2.4. Vitamin D and Ergosterol

Vitamin D deficiency is a global concern for most vegetarians, as most vitamin D-rich foods are derived from animals. Under ultraviolet irradiation, ergosterol is photolyzed to produce a vitamin D precursor, and then slowly isomerized to vitamin D by a thermal reaction. Ergosterol is an important part of the fungal cell membrane, and its chemical structure is similar to cholesterol. It has been observed that ergosterol can enhance the cell membrane of mushrooms, regulate membrane fluidity, and assist membrane transportation, which is similar to animal cholesterol [55]. The free and esterified forms of ergosterol play an important role in membrane fluidity and integrity. Therefore, ergosterol is a target for a variety of antifungal drugs [3].

According to the previous research results, which showed that the vitamin D content in the cultivated *A. bisporus* is very low [56]. Generally, when air or oxygen was used as the feed gas, ozone was a key substance for the efficiency of cold plasma. At this time, the ultraviolet radiation power generated by cold plasma was very small, making it impossible for ergosterol in mushrooms to be converted into vitamin D through ultraviolet radiation. Although, the reactive oxygen species (ROS) and reactive nitrogen species (RNS) produced by cold plasma could cause cell damage, but this was not the same as the effects of ultraviolet radiation [47,57,58]. we did not find the presence of vitamin D in *A. bisporus* that was treated with cold plasma, which might explain the research results.

In Figure 5E, it is observed that the ergosterol content of *A.bisporus* in the CP and control groups follows a gradually increasing trend as the storage period progresses. On the last day of storage, it is significantly higher than in the other two treatment groups ($p < 0.05$). Since the mushrooms used in this experiment were freshly harvested, the mushrooms still have biological activity and metabolic capacity. Mushrooms are known for their high metabolic capacity and growth rate; induction of ergosterol synthesis following prolonged exposed storage may have occurred [59].

On the third day of storage, the ergosterol content in *A. bisporus* treated with cold plasma was higher than that in the PE group ($p < 0.05$). However, on the sixth day, the ergosterol content in the PE group increased by about 65%. This was different from the study of traditional cold plasma treatment of fungi and also different from the results obtained in the treatment of non-edible fungi with cold plasma. *A. bisporus* is a large fungi and tends to have more complex cell wall structures, which leads to this difference [60, 61]. The content of ergosterol in CP+PE group did not increase, which may be due to the oxidative stress on the mushroom cell membrane caused by cold plasma treatment, resulting in the ergosterol participating in antioxidant metabolism. At the later stage of storage, ROS and RNS in the packaging group decreased, and there was no significant difference in ergosterol content in the packaging group ($p > 0.05$).

3.2.5. Correlation Analysis and PLS-DA

The biochemical characteristics and sensory characteristics of *A. bisporus* undergo different degrees of change during storage. Figure 6A shows the correlation analysis of the indicators of *A. bisporus* between different treatment groups. Storage time is significantly correlated with browning degree, hardness, weight loss rate, total number of colonies, and soluble solids. The changes of L* and a* are also significantly correlated with total phenol and total protein. However, no significant correlation was found with the changes in PPO and POD enzymes, which indicates that cold plasma treatment regulates the changes in the factors leading to browning. Figure 6B shows the principal component analysis of the effects of different treatment methods on the quality of *A. bisporus* during storage. The pre-member rates of the first and second principal components were 46.8% and 22.2%, respectively. During the storage of *A. bisporus*, the soluble solids content, weight loss rate, browning degree, and total number of colonies contributed more to the first and second principal components. Furthermore, ergosterol, total phenol, total protein, and the water content change index contributed more to the first principal component, while hardness and vitamin C contributed more to the second principal component. Therefore, soluble solids content, weight loss rate, browning degree, and the total number of colonies were the key indicators during the storage of *A. bisporus*. With an increase in storage time, the CP and control groups moved to the first quadrant due to the lack of packaging protection, which was represented by the serious water loss and browning of the mushrooms. The PE group moved to the lower half of the storage period until it appeared in the fourth quadrant, indicating that the *A. bisporus* was infected with microorganisms and accompanied by browning at this time. The CP+PE group represented better quality, which manifested as two aspects. Firstly, there was no serious browning and water loss. Due to the bactericidal effect of the reactive oxygen species, there was no large-scale microbial infection in the packaging. On the other hand, the nutrients in *A. bisporus* did not suffer serious losses and remained relatively stable. This was similar to the results of the correlation analysis, which proved that the cold plasma treatment in the packaging affected the storage process of *A. bisporus* by increasing the sterility and reducing water loss.

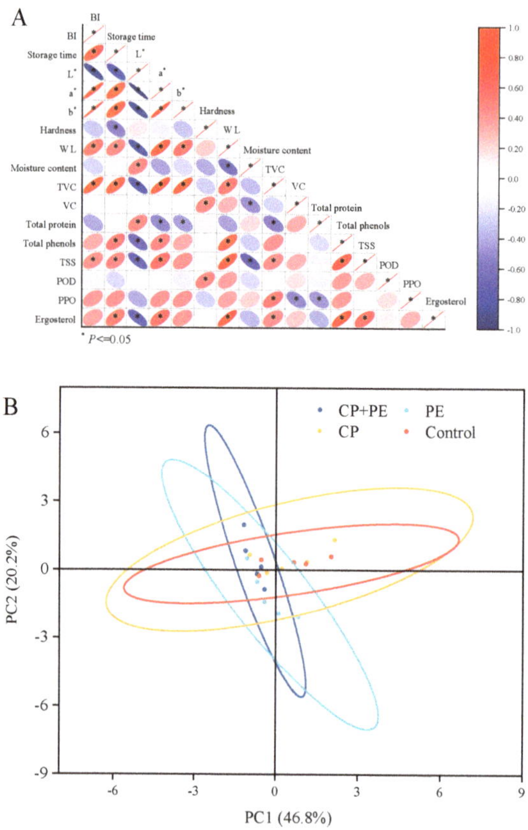

Figure 6. The correlation analysis (**A**) and principal component analysis (**B**).

4. Conclusions

This study determined the optimal conditions for cold plasma treatment (95 kV, 130 Hz, 10 min) and evaluated its impact on *A. bisporus* quality during storage. The CP+PE group effectively reduced the presence of spoilage microorganisms and minimized the risk of contamination post-treatment, while simultaneously preserving key quality characteristics such as hardness, weight loss rate, moisture content, and browning. Furthermore, the treatment minimized the loss of total protein and TSS in postharvest *A. bisporus*. In contrast, the PE packaging group exhibited severe browning, higher microbial growth, and lower vitamin C content during storage. Interestingly, the treatment reduced the activity of the PPO enzyme but stimulated the activity of POD, which contributed to maintaining the color of *A. bisporus*. These results were further supported by correlation and principal component analyses. Overall, this study highlighted the promising application of optimized cold plasma technology in eliminating pathogenic microorganisms and ensuring stable quality in fresh agricultural products. It also provided a valuable research foundation for the use of cold plasma technology in the storage and preservation of such products.

Supplementary Materials: The following supporting information can be downloaded at: https://www.mdpi.com/article/10.3390/foods13213393/s1, Figure S1. Schematic diagram of Agaricus bisporus storage box. Figure S2. Effects of different treatment voltages on the whiteness (A), total number of colonies and sterilization rate (B), appearance (C) of *Agaricus bisporus*. Figure S3. Effects of different treatment frequencies on the whiteness (A), total number of colonies and sterilization rate (B), appearance (C) of *Agaricus bisporus*. Figure S4. Effects of different processing time on the whiteness (A), total number of colonies and sterilization rate (B), appearance (C) of *Agaricus bisporus*. Table S1. Experiment design and results for response surface analysis.

Author Contributions: Y.G.: Conceptualization, Data curation, Validation, Writing—original draft, Writing—review and editing, Supervision, Project administration. S.X.: Validation, Investigation, Resources. C.S.: Validation, Investigation, Resources. N.M.: Validation, Resources. F.P.: Validation, Resources. W.Y.: Validation, Resources. Q.H.: Validation, Resources. B.M.K.: Validation, Review and editing. D.F.: Writing—review and editing, Investigation, Supervision, Project administration, Funding acquisition. All authors have read and agreed to the published version of the manuscript.

Funding: This study was funded by the Outstanding Youth Fund Project of Natural Science Foundation of Jiangsu Province (BK20240140), the Basic Science (Natural Science) Research Project of Colleges and Universities in Jiangsu Province (23KJB550009), the Jiangsu Association for Science & Technology Youth Science & Technology Talents Lifting Project (JSTJ-2023-XH027) and Qing Lan Project of Jiangsu.

Institutional Review Board Statement: Not applicable.

Informed Consent Statement: Not applicable.

Data Availability Statement: The original contributions presented in the study are included in the article/Supplementary Materials, further inquiries can be directed to the corresponding author.

Conflicts of Interest: There are no conflicts of interest to declare.

References

1. Ramos, M.; Burgos, N.; Barnard, A.; Evans, G.; Preece, J.; Graz, M.; Ruthes, A.C.; Jiménez-Quero, A.; Martínez-Abad, A.; Vilaplana, F.; et al. *Agaricus bisporus* and Its By-Products as a Source of Valuable Extracts and Bioactive Compounds. *Food Chem.* **2019**, *292*, 176–187. [CrossRef]
2. Blumfield, M.; Abbott, K.; Duve, E.; Cassettari, T.; Marshall, S.; Fayet-Moore, F. Examining the Health Effects and Bioactive Components in *Agaricus bisporus* Mushrooms: A Scoping Review. *J. Nutr. Biochem.* **2020**, *84*, 108453. [CrossRef]
3. Zhang, K.; Pu, Y.-Y.; Sun, D.-W. Recent Advances in Quality Preservation of Postharvest Mushrooms (*Agaricus bisporus*): A Review. *Trends Food Sci. Technol.* **2018**, *78*, 72–82. [CrossRef]
4. Sun, T.; Bian, J.; Wang, Y.; Hu, J.; Yun, X.; Chen, E.; Dong, T. One-Step Synthesis of Poly(L-Lactic Acid)-Based Soft Films with Gas Permselectivity for White Mushrooms (*Agaricus bisporus*) Preservation. *Foods* **2023**, *12*, 586. [CrossRef]
5. Yan, X.; Cheng, M.; Wang, Y.; Zhao, P.; Wang, K.; Wang, Y.; Wang, X.; Wang, J. Evaluation of Film Packaging Containing Mesoporous Nanosilica and Oregano Essential Oil for Postharvest Preservation of Mushrooms (*Agaricus bisporus*). *Postharvest Biol. Technol.* **2023**, *198*, 112263. [CrossRef]
6. Zheng, B.; Kou, X.; Liu, C.; Wang, Y.; Yu, Y.; Ma, J.; Liu, Y.; Xue, Z. Effect of Nanopackaging on the Quality of Edible Mushrooms and Its Action Mechanism: A Review. *Food Chem.* **2023**, *407*, 135099. [CrossRef]
7. Feng, Y.; Xu, H.; Sun, Y.; Xia, R.; Hou, Z.; Li, Y.; Wang, Y.; Pan, S.; Li, L.; Zhao, C.; et al. Effect of Light on Quality of Preharvest and Postharvest Edible Mushrooms and Its Action Mechanism: A Review. *Trends Food Sci. Technol.* **2023**, *139*, 104119. [CrossRef]
8. Dong, S.; Guo, J.; Yu, J.; Bai, J.; Xu, H.; Li, M. Effects of Electron-Beam Generated X-Ray Irradiation on the Postharvest Storage Quality of *Agaricus bisporus*. *Innov. Food Sci. Emerg. Technol.* **2022**, *80*, 103079. [CrossRef]
9. Bezerra, J.d.A.; Lamarao, C.V.; Sanches, E.A.; Rodrigues, S.; Fernandes, F.A.N.; Ramos, G.L.P.A.; Esmerino, E.A.; Cruz, A.G.; Campelo, P.H. Cold Plasma as a Pre-Treatment for Processing Improvement in Food: A Review. *Food Res. Int.* **2023**, *167*, 112663. [CrossRef]
10. Farooq, S.; Dar, A.H.; Dash, K.K.; Srivastava, S.; Pandey, V.K.; Ayoub, W.S.; Pandiselvam, R.; Manzoor, S.; Kaur, M. Cold Plasma Treatment Advancements in Food Processing and Impact on the Physiochemical Characteristics of Food Products. *Food Sci. Biotechnol.* **2023**, *32*, 621–638. [CrossRef]
11. Zhou, R.; Rezaeimotlagh, A.; Zhou, R.; Zhang, T.; Wang, P.; Hong, J.; Soltani, B.; Mai-Prochnow, A.; Liao, X.; Ding, T.; et al. In-Package Plasma: From Reactive Chemistry to Innovative Food Preservation Technologies. *Trends Food Sci. Technol.* **2022**, *120*, 59–74. [CrossRef]
12. Ding, Y.; Mo, W.; Deng, Z.; Kimatu, B.M.; Gao, J.; Fang, D. Storage Quality Variation of Mushrooms (*Flammulina velutipes*) after Cold Plasma Treatment. *Life* **2023**, *13*, 70. [CrossRef] [PubMed]
13. Subrahmanyam, K.; Gul, K.; Sehrawat, R.; Allai, F.M. Impact of In-Package Cold Plasma Treatment on the Physicochemical Properties and Shelf Life of Button Mushrooms (*Agaricus bisporus*). *Food Biosci.* **2023**, *52*, 102425. [CrossRef]
14. Jia, S.; Zhang, N.; Dong, C.; Zheng, P.; Ji, H.; Yu, J.; Yan, S.; Chen, C.; Liang, L. Effect of Cold Plasma Treatment on the Softening of Winter Jujubes (*Ziziphus jujuba* Mill. Cv. Dongzao). *Horticulturae* **2023**, *9*, 986. [CrossRef]
15. Hua, X.; Li, T.; Wu, C.; Zhou, D.; Fan, G.; Li, X.; Cong, K.; Yan, Z.; Wu, Z. Novel Physical Treatments (Pulsed Light and Cold Plasma) Improve the Quality of Postharvest Apricots after Long-Distance Simulated Transportation. *Postharvest Biol. Technol.* **2022**, *194*, 112098. [CrossRef]

16. Du, Y.; Mi, S.; Wang, H.; Yuan, S.; Yang, F.; Yu, H.; Xie, Y.; Guo, Y.; Cheng, Y.; Yao, W. Intervention Mechanisms of Cold Plasma Pretreatment on the Quality, Antioxidants and Reactive Oxygen Metabolism of Fresh Wolfberries during Storage. *Food Chem.* **2024**, *431*, 137106. [CrossRef]
17. Dharini, M.; Jaspin, S.; Mahendran, R. Cold Plasma Reactive Species: Generation, Properties, and Interaction with Food Biomolecules. *Food Chem.* **2023**, *405*, 134746. [CrossRef] [PubMed]
18. Waghmare, R. Cold Plasma Technology for Fruit Based Beverages: A Review. *Trends Food Sci. Technol.* **2021**, *114*, 60–69. [CrossRef]
19. Zhu, Y.; Elliot, M.; Zheng, Y.; Chen, J.; Chen, D.; Deng, S. Aggregation and Conformational Change of Mushroom (*Agaricus bisporus*) Polyphenol Oxidase Subjected to Atmospheric Cold Plasma Treatment. *Food Chem.* **2022**, *386*, 132707. [CrossRef]
20. Mahdavian Mehr, H.; Koocheki, A. Effects of Short-Term and Long-Term Cold Plasma Treatment on the Color, Structure, and Pickering Foaming Properties of Grass Pea Protein Particles. *Food Hydrocoll.* **2023**, *143*, 108846. [CrossRef]
21. Ke, Z.; Bai, Y.; Zhu, H.; Xiang, X.; Liu, S.; Zhou, X.; Ding, Y. Characteristics of Myoglobin Degradation by Cold Plasma and Its Pro-Oxidative Activity on Lipid in Washed Fish Muscle. *Food Chem.* **2022**, *389*, 132972. [CrossRef] [PubMed]
22. Yang, X.; Cheng, J.-H.; Sun, D.-W. Enhancing Microorganism Inactivation Performance through Optimization of Plate-to-Plate Dielectric Barrier Discharge Cold Plasma Reactors. *Food Control* **2024**, *157*, 110164. [CrossRef]
23. Bourke, P.; Ziuzina, D.; Boehm, D.; Cullen, P.J.; Keener, K. The Potential of Cold Plasma for Safe and Sustainable Food Production. *Trends Biotechnol.* **2018**, *36*, 615–626. [CrossRef]
24. Yan, M.; Yuan, B.; Xie, Y.; Cheng, S.; Huang, H.; Zhang, W.; Chen, J.; Cao, C. Improvement of Postharvest Quality, Enzymes Activity and Polyphenoloxidase Structure of Postharvest *Agaricus bisporus* in Response to High Voltage Electric Field. *Postharvest Biol. Technol.* **2020**, *166*, 111230. [CrossRef]
25. Shi, C.; Zhou, A.; Fang, D.; Lu, T.; Wang, J.; Song, Y.; Lyu, L.; Wu, W.; Huang, C.; Li, W. Oregano Essential Oil/β-Cyclodextrin Inclusion Compound Polylactic Acid/Polycaprolactone Electrospun Nanofibers for Active Food Packaging. *Chem. Eng. J.* **2022**, *445*, 136746. [CrossRef]
26. Fattahifar, E.; Barzegar, M.; Ahmadi Gavlighi, H.; Sahari, M.A. Evaluation of the Inhibitory Effect of Pistachio (*Pistacia vera* L.) Green Hull Aqueous Extract on Mushroom Tyrosinase Activity and Its Application as a Button Mushroom Postharvest Anti-Browning Agent. *Postharvest Biol. Technol.* **2018**, *145*, 157–165. [CrossRef]
27. Ni, X.; Yu, J.; Shao, P.; Yu, J.; Chen, H.; Gao, H. Preservation of *Agaricus bisporus* Freshness with Using Innovative Ethylene Manipulating Active Packaging Paper. *Food Chem.* **2021**, *345*, 128757. [CrossRef]
28. Sun, Y.; Nzekoue, F.K.; Vittori, S.; Sagratini, G.; Caprioli, G. Conversion of Ergosterol into Vitamin D2 and Other Photoisomers in *Agaricus Bisporus* Mushrooms under UV-C Irradiation. *Food Biosci.* **2022**, *50*, 102143. [CrossRef]
29. Du, Y.; Yang, F.; Yu, H.; Xie, Y.; Yao, W. Improving Food Drying Performance by Cold Plasma Pretreatment: A Systematic Review. *Compr. Rev. Food Sci. Food Saf.* **2022**, *21*, 4402–4421. [CrossRef]
30. Shanker, M.A.; Khanashyam, A.C.; Pandiselvam, R.; Joshi, T.J.; Thomas, P.E.; Zhang, Y.; Rustagi, S.; Bharti, S.; Thirumdas, R.; Kumar, M.; et al. Implications of Cold Plasma and Plasma Activated Water on Food Texture—A Review. *Food Control* **2023**, *151*, 109793. [CrossRef]
31. Li, L.; Kitazawa, H.; Zhang, X.; Zhang, L.; Sun, Y.; Wang, X.; Liu, Z.; Guo, Y.; Yu, S. Melatonin Retards Senescence via Regulation of the Electron Leakage of Postharvest White Mushroom (*Agaricus bisporus*). *Food Chem.* **2021**, *340*, 127833. [CrossRef]
32. Sun, B.; Chen, X.; Xin, G.; Qin, S.; Chen, M.; Jiang, F. Effect of 1-Methylcyclopropene (1-MCP) on Quality of Button Mushrooms (*Agaricus bisporus*) Packaged in Different Packaging Materials. *Postharvest Biol. Technol.* **2020**, *159*, 111023. [CrossRef]
33. Qin, Y.; Liu, D.; Wu, Y.; Yuan, M.; Li, L.; Yang, J. Effect of PLA/PCL/Cinnamaldehyde Antimicrobial Packaging on Physicochemical and Microbial Quality of Button Mushroom (*Agaricus bisporus*). *Postharvest Biol. Technol.* **2015**, *99*, 73–79. [CrossRef]
34. Pourbagher, R.; Abbaspour-Fard, M.H.; Sohbatzadeh, F.; Rohani, A. Inhibition of Enzymes and *Pseudomonas tolaasii* Growth on *Agaricus bisporus* Following Treatment with Surface Dielectric Barrier Discharge Plasma. *Innov. Food Sci. Emerg. Technol.* **2021**, *74*, 102833. [CrossRef]
35. Yarabbi, H.; Soltani, K.; Sangatash, M.M.; Yavarmanesh, M.; Shafafi Zenoozian, M. Reduction of Microbial Population of Fresh Vegetables (Carrot, White Radish) and Dried Fruits (Dried Fig, Dried Peach) Using Atmospheric Cold Plasma and Its Effect on Physicochemical Properties. *J. Agric. Food Res.* **2023**, *14*, 100789. [CrossRef]
36. Lin, X.; Sun, D.-W. Research Advances in Browning of Button Mushroom (*Agaricus bisporus*): Affecting Factors and Controlling Methods. *Trends Food Sci. Technol.* **2019**, *90*, 63–75. [CrossRef]
37. Zhao, Z.; Wang, X.; Ma, T. Properties of Plasma-Activated Water with Different Activation Time and Its Effects on the Quality of Button Mushrooms (*Agaricus bisporus*). *LWT-FOOD Sci. Technol.* **2021**, *147*, 111633. [CrossRef]
38. Wei, W.; Lv, P.; Xia, Q.; Tan, F.; Sun, F.; Yu, W.; Jia, L.; Cheng, J. Fresh-Keeping Effects of Three Types of Modified Atmosphere Packaging of Pine-Mushrooms. *Postharvest Biol. Technol.* **2017**, *132*, 62–70. [CrossRef]
39. Sruthi, N.U.; Josna, K.; Pandiselvam, R.; Kothakota, A.; Gavahian, M.; Mousavi Khaneghah, A. Impacts of Cold Plasma Treatment on Physicochemical, Functional, Bioactive, Textural, and Sensory Attributes of Food: A Comprehensive Review. *Food Chem.* **2022**, *368*, 130809. [CrossRef]
40. Dong, S.; Fan, L.; Ma, Y.; Du, J.; Xiang, Q. Inactivation of Polyphenol Oxidase by Dielectric Barrier Discharge (DBD) Plasma: Kinetics and Mechanisms. *LWT* **2021**, *145*, 111322. [CrossRef]

41. Pourbagher, R.; Abbaspour-Fard, M.H.; Khomeiri, M.; Sohbatzadeh, F.; Rohani, A. Effects of Gas Type and Cold Plasma Treatment Time on Lecanicillium Fungicola Spores Reduction and Changes in Qualitative, Chemical, and Physiological Characteristics of Button Mushroom during Postharvest Storage. *J. Food Process. Preserv.* **2022**, *46*, e16901. [CrossRef]
42. Shen, C.; Chen, W.; Li, C.; Cui, H.; Lin, L. The Effects of Cold Plasma (CP) Treatment on the Inactivation of Yam Peroxidase and Characteristics of Yam Slices. *J. Food Eng.* **2023**, *359*, 111693. [CrossRef]
43. Dantas, A.M.; Batista, J.D.F.; dos Santos Lima, M.; Fernandes, F.A.N.; Rodrigues, S.; Magnani, M.; Borges, G.d.S.C. Effect of Cold Plasma on Açai Pulp: Enzymatic Activity, Color and Bioaccessibility of Phenolic Compounds. *LWT* **2021**, *149*, 111883. [CrossRef]
44. Zhou, D.; Wang, Z.; Tu, S.; Chen, S.; Peng, J.; Tu, K. Effects of Cold Plasma, UV-C or Aqueous Ozone Treatment on Botrytis Cinerea and Their Potential Application in Preserving Blueberry. *J. Appl. Microbiol.* **2019**, *127*, 175–185. [CrossRef]
45. Mayookha, V.P.; Pandiselvam, R.; Kothakota, A.; Padma Ishwarya, S.; Chandra Khanashyam, A.; Kutlu, N.; Rifna, E.J.; Kumar, M.; Panesar, P.S.; Abd El-Maksoud, A.A. Ozone and Cold Plasma: Emerging Oxidation Technologies for Inactivation of Enzymes in Fruits, Vegetables, and Fruit Juices. *Food Control* **2023**, *144*, 109399. [CrossRef]
46. Wang, L.; Chen, Q.; Zhang, J.; Cheng, P.; Hu, J.; Dong, T. Effect of Modified Atmosphere Packaging Materials on Physicochemical and Selected Enzyme Activities of Agaricus Bernardii. *J. Food Process Eng.* **2021**, *44*, e13628. [CrossRef]
47. Leite, A.K.F.; Fonteles, T.V.; Miguel, T.B.A.R.; Silvestre da Silva, G.; Sousa de Brito, E.; Alves Filho, E.G.; Fernandes, F.A.N.; Rodrigues, S. Atmospheric Cold Plasma Frequency Imparts Changes on Cashew Apple Juice Composition and Improves Vitamin C Bioaccessibility. *Food Res. Int.* **2021**, *147*, 110479. [CrossRef]
48. Khoshkalam Pour, A.; Khorram, S.; Ehsani, A.; Ostadrahimi, A.; Ghasempour, Z. Atmospheric Cold Plasma Effect on Quality Attributes of Banana Slices: Its Potential Use in Blanching Process. *Innov. Food Sci. Emerg. Technol.* **2022**, *76*, 102945. [CrossRef]
49. Wang, X.-M.; Zhang, J.; Wu, L.-H.; Zhao, Y.-L.; Li, T.; Li, J.-Q.; Wang, Y.-Z.; Liu, H.-G. A Mini-Review of Chemical Composition and Nutritional Value of Edible Wild-Grown Mushroom from China. *Food Chem.* **2014**, *151*, 279–285. [CrossRef]
50. Billah, M.; Sajib, S.A.; Roy, N.C.; Rashid, M.M.; Reza, M.A.; Hasan, M.M.; Talukder, M.R. Effects of DBD Air Plasma Treatment on the Enhancement of Black Gram (Vigna Mungo l.) Seed Germination and Growth. *Arch. Biochem. Biophys.* **2020**, *681*, 108253. [CrossRef]
51. Ji, W.; Li, M.; Yang, T.; Li, H.; Li, W.; Wang, J.; Ma, M. Effect of Cold Plasma on Physical–Biochemical Properties and Nutritional Components of Soybean Sprouts. *Food Res. Int.* **2022**, *161*, 111766. [CrossRef]
52. Bahrami, N.; Bayliss, D.; Chope, G.; Penson, S.; Perehinec, T.; Fisk, I.D. Cold Plasma: A New Technology to Modify Wheat Flour Functionality. *Food Chem.* **2016**, *202*, 247–253. [CrossRef]
53. Dong, J.; Zhang, M.; Lu, L.; Sun, L.; Xu, M. Nitric Oxide Fumigation Stimulates Flavonoid and Phenolic Accumulation and Enhances Antioxidant Activity of Mushroom. *Food Chem.* **2012**, *135*, 1220–1225. [CrossRef]
54. Guo, Y.; Guo, S.; Li, M.; Zhang, R.; Liu, Z.; Wang, X. The Effects of LTP/CEO/SBA-15 Potato Starch Film on the Postharvest Quality of *Agaricus bisporus*. *Sci. Hortic.* **2024**, *324*, 112576. [CrossRef]
55. Cardwell, G.; Bornman, J.F.; James, A.P.; Black, L.J. A Review of Mushrooms as a Potential Source of Dietary Vitamin D. *Nutrients* **2018**, *10*, 1498. [CrossRef]
56. Shao, S.; Hernandez, M.; Kramer, J.K.G.; Rinker, D.L.; Tsao, R. Ergosterol Profiles, Fatty Acid Composition, and Antioxidant Activities of Button Mushrooms as Affected by Tissue Part and Developmental Stage. *J. Agric. Food Chem.* **2010**, *58*, 11616–11625. [CrossRef]
57. Feizollahi, E.; Misra, N.N.; Roopesh, M.S. Factors Influencing the Antimicrobial Efficacy of Dielectric Barrier Discharge (DBD) Atmospheric Cold Plasma (ACP) in Food Processing Applications. *Crit. Rev. Food Sci. Nutr.* **2021**, *61*, 666–689. [CrossRef]
58. Wang, J.; Fu, T.; Sang, X.; Liu, Y. Effects of High Voltage Atmospheric Cold Plasma Treatment on Microbial Diversity of Tilapia (Oreochromis Mossambicus) Fillets Treated during Refrigeration. *Int. J. Food Microbiol.* **2022**, *375*, 109738. [CrossRef]
59. Simon, R.R.; Phillips, K.M.; Horst, R.L.; Munro, I.C. Vitamin D Mushrooms: Comparison of the Composition of Button Mushrooms (*Agaricus bisporus*) Treated Postharvest with UVB Light or Sunlight. *J. Agric. Food Chem.* **2011**, *59*, 8724–8732. [CrossRef] [PubMed]
60. Rahimi-Verki, N.; Shapoorzadeh, A.; Razzaghi-Abyaneh, M.; Atyabi, S.-M.; Shams-Ghahfarokhi, M.; Jahanshiri, Z.; Gholami-Shabani, M. Cold Atmospheric Plasma Inhibits the Growth of Candida Albicans by Affecting Ergosterol Biosynthesis and Suppresses the Fungal Virulence Factors in Vitro. *Photodiagnosis Photodyn. Ther.* **2016**, *13*, 66–72. [CrossRef]
61. Shapourzadeh, A.; Rahimi-Verki, N.; Atyabi, S.-M.; Shams-Ghahfarokhi, M.; Jahanshiri, Z.; Irani, S.; Razzaghi-Abyaneh, M. Inhibitory Effects of Cold Atmospheric Plasma on the Growth, Ergosterol Biosynthesis, and Keratinase Activity in Trichophyton Rubrum. *Arch. Biochem. Biophys.* **2016**, *608*, 27–33. [CrossRef]

foods

MDPI

Article

Effect of High CO_2 Controlled Atmosphere Storage on Postharvest Quality of Button Mushroom (*Agaricus bisporus*)

Yuxian Yang [1], Ouyang Jia [1], Yunzhi Li [1], Bing Feng [1], Mingchang Chang [1,2], Junlong Meng [1,*] and Bing Deng [1,2,*]

[1] College of Food Science and Engineering, Shanxi Agricultural University, Jinzhong 030801, China; soph0731@163.com (Y.Y.); m19834542144@163.com (O.J.); 18214659894@163.com (Y.L.); 15738227612@163.com (B.F.); sxndcmc@163.com (M.C.)
[2] Shanxi Key Laboratory of Edible Fungi for Loess Plateau, Jinzhong 030801, China
* Correspondence: mengjunlongseth@126.com (J.M.); dengbing_edu@163.com (B.D.)

Abstract: The *Agaricus bisporus* (Button mushroom) stands out as one of the most prolific edible fungi which offers robust flavor and nutrition. Nonetheless, this mushroom contains high moisture levels and intense respiration. Without appropriate postharvest preservation techniques, the button mushroom readily experiences browning and senescence. To ensure optimum quality, prompt cooling and appropriate storage conditions are essential. This present research investigated the postharvest quality of button mushrooms stored in a controlled atmosphere (CA) with different initial gas compositions. The findings revealed that button mushrooms in the CA group demonstrated considerable enhancements in appearance and overall quality, effectively delaying browning and senescence compared to those in the control group. The optimal gas composition is 1–3% O_2 and 15–17% CO_2 (CAII), which effectively inhibited the expression of polyphenol oxidase (PPO)- and lactase (LAC)-related genes in the button mushroom, maintaining a high L* value. Furthermore, the application of 1–3% O_2 and 15–17% CO_2 (CAII) not only preserved visual quality but also extended the postharvest shelf life of the button mushroom by minimizing metabolic activities that contribute to senescence. Moreover, 1–3% O_2 and 15–17% CO_2 (CAII) storage also reduced the expression levels of genes associated with ethylene synthesis, which is reflected in the gradual decrease in cell membrane permeability. Consequently, this research underscores the critical importance of controlled atmosphere storage in improving the marketability and sustainability of this widely consumed mushroom.

Keywords: *Agaricus bisporus*; controlled atmosphere; enzymatic browning; senescence; gene expression

1. Introduction

Agaricus bisporus, commonly referred to as the button mushroom or white mushroom, stands among the most widely cultivated edible fungi worldwide with a delightful flavor and notable nutritional advantages [1,2]. However, the fruiting bodies of the button mushroom harbor high moisture content and exhibit elevated respiration rates. Consequently, they rapidly consume vital nutrients and water, which are crucial for maintaining normal physiological processes [3–5]. Moreover, the absence of distinct protective features on the surface of the button mushroom makes it susceptible to external physical damage and microbial infections, consequently resulting in browning and senescence [6]. To satisfy market needs, efficient postharvest storage techniques must be employed to prolong the shelf life of button mushrooms. Additionally, implementing advanced methods such as irradiation, ultrasound, essential oils, salicylic acid, citric acid, and methyl jasmonate can effectively reduce postharvest losses in fruits and vegetables [7–12]. Nevertheless, these approaches face limitations, including high expenses, safety risks, nutrient degradation, and changes in texture. Conversely, to extend the shelf life of fresh-cut fruits, vegetables, and edible fungi, researchers have studied and implemented a controlled atmosphere (CA) to inhibit microorganisms and maintain product quality. Initially, CA applications typically included reduced O_2 and increased CO_2 levels. This adjustment not only enhances fruit and vegetable quality but also inhibits microbial growth, contributing to a more sustainable supply chain [13]. As an illustration, CA storage significantly reduces browning in fresh in-hull walnuts and increases their postharvest lifespan by over 1.6 times [14]. Moreover, it has been proven that CO_2-rich modified atmospheres influence energy metabolism, respiration rates, ethylene responses, and physiological changes in various fresh products during packaging and postharvest storage [15]. Research by Belay, Z. A. demonstrates that low O_2 (1–5 kPa) and high CO_2 (5–15 kPa) atmospheres can extend the life span of fresh-cut fruits by reducing respiration, inhibiting ethylene biosynthesis, and suppressing the growth of aerobic microorganisms [16]. Whereas, compared to horticultural products, edible fungi exhibit unique traits such as intense postharvest respiration and significant tolerance to external CO_2 levels, suggesting that the air conditioning parameters of fresh edible mushrooms represented by the button mushroom may differ from those of fruits and vegetables. In a previous study, a short time exposure of button mushrooms to a high CO_2 concentration resulted in reduced browning and maintained flavor quality, which means that, in contrast to the low CO_2 atmosphere typically used in fruits and vegetables, high CO_2 concentrations may help delay the senescence of button mushrooms and prolong their postharvest life [17]. Additionally, Amodio et al. demonstrated that a CA composition of 3% O_2, 20% CO_2, and 77% N_2 effectively slows down browning and firmness loss in fresh-cut *Pleurotus eryngii* while inhibiting fungal growth [18].

Browning has the most pronounced effect on the visual appeal of edible fungi. Mechanically, browning is divided into two categories: enzymatic and non-enzymatic [19]. Enzymatic browning poses a significant challenge for horticultural products like fruits (e.g., apples, peaches, pears, bananas, and avocados) and vegetables (e.g., lettuce, potatoes, and eggplants) [20]. Recent studies indicate that the browning observed in edible fungi, particularly button mushrooms, primarily results from enzymatic processes, with polyphenol oxidase (PPO) being the chief enzyme involved [21]. In this mechanism, phenolic compounds oxidize and form quinones, which subsequently react with amino acids, leading to melanin production and resultant browning [22]. Recent findings have shown that during the storage of button mushrooms, the L* value decreases, reflecting a rise in PPO enzyme activity along with an elevated expression of the *PPO3* and *PPO4* genes [23].

Senescence is also one of the important factors that affect the shelf life of edible fungi. Research reveals that ethylene release plays a critical role in determining their storage longevity. Ethylene, identified as a plant hormone, acts as a signaling molecule that affects several metabolic pathways. As the button mushroom matures, its ethylene production may hasten senescence, resulting in changes to texture, color, and nutrient profile. Furthermore, several studies emphasize the crucial role of ethylene in the senescence of straw-rotting fungi, including the button mushroom, straw mushroom, and oyster mushroom, through the activation of genes linked to cell wall degradation, pigment accumulation, and aromatic compound production [24–26]. Notably, during button mushroom senescence, genes responsible for ethylene biosynthesis (*SAMS*, *ACS*, and *ACO*) play crucial roles in the postharvest management of button mushrooms.

This investigation seeks to evaluate how controlled atmosphere storage, using various initial gas compositions, impacts the visual appearance and quality characteristics of fresh button mushrooms during storage. Specifically, we focus on key indicators of freshness, conducting thorough testing and analysis on hardness, weight loss, color, and the expression of enzymatic browning-related and ethylene synthesis-related genes. Additionally, further exploration into specific storage parameters that affect browning and senescence may offer valuable insights for both consumers and producers alike.

2. Materials and Methods

2.1. Preparation of A. bisporus and Storage

The newly harvested button mushrooms were sourced from Yisheng Microbiology Co., Ltd., Lin Fen, China. On the day of harvesting, samples of button mushroom were taken to the laboratory and the experiments exclusively selected button mushrooms of superior quality, featuring sealed caps, and devoid of any imperfections. Thereafter, fresh button mushrooms were stored under each of the following conditions: (1) 1–3% O_2 and 13–15% CO_2 (CAI); (2) 1–3% O_2 and 15–17% CO_2 (CAII); (3) 3–5% O_2 and 13–15% CO_2 (CAIII); (4) 3–5% O_2 and 15–17% CO_2 (CAIV); and (5) ambient air (control). Initially, all samples underwent pre-cooling for 12 h within a cold room maintained at 4 ± 1 °C. Due to the respiration of button mushrooms affecting O_2 and CO_2 levels in controlled atmosphere chambers, daily adjustments were made to maintain the initial set levels of O_2 and CO_2 values. The samples were taken at three-day intervals. During storage, certain samples of button mushrooms were preserved for subsequent index analysis.

2.2. Experimental Methods

2.2.1. Measurement of Weight Loss

Weight was measured on every sampling day. Three parallel experiments were conducted during each measurement, with each experimental group containing 10 mushroom samples.

2.2.2. Measurement of Hardness

The texture of button mushrooms was analyzed using a TMS-PRO (FTC, CA, USA) texture analyzer. First, the stems of the button mushrooms were removed and a 2 mm thick section of the cap was cut [27]. Measurements were taken at the center of the cap and two equidistant positions on either side, making a total of 3 measurement points. The experimental conditions were as follows: testing speed: 100 mm/min, force-sensing range: 100 N, probe diameter: 0.5 cm, and distance: 8 mm. Each treatment group tested 9 mushroom samples and conducted three repeated measurements, taking the average of the results. Samples from each of the treatment groups were tested every three days.

2.2.3. Measurement of Visual Appearance and Color

A digital camera (Nikon D7500, Nikon Corporation, Shanghai, China) was used to record the surface and internal appearance of button mushroom. Photos were taken of three mushrooms for each treatment every three days.

The L* value represents the brightness of button mushrooms, with higher values indicating whiter or less browning on the button mushroom surface. Changes in the color of the button mushrooms (L* value) were periodically measured using a colorimeter (WSC-S, Shanghai, China). The measurement method involved randomly selecting nine button mushrooms from each experimental group and fixing three points on the surface of the button mushroom surface. Additionally, nine button mushrooms were cut in the middle from the cap to the stem, and the color changes within the button mushrooms were measured. Measurements were taken every 3 days and three parallel experiments were conducted during each measurement, with each group containing 10 mushroom samples. The results were represented as the average.

2.2.4. Measurement of Membrane Permeability

The relative membrane conductivity of the button mushroom samples was determined according to the method in accordance with Wantat, Seraypheap, and Rojsitthisak with some modifications [28]. During the experiment, 6 mushrooms were selected from each treatment group every 3 days for sampling. A perforator was employed to create openings in the mushrooms, subsequently positioning them in glass beakers filled with 20 mL of distilled water. The specimens were left undisturbed for half an hour, after which their electrical conductivity was gauged using a conductivity meter (DDS-11A, Shanghai, China), labeled P1. To eliminate mushroom tissues, the samples underwent a 10 min heating in boiling water. Following a 10 min period of cooling to ambient temperature, the electrical conductivity was reassessed, labeled as P2. The procedure was replicated three times, followed by the computation of the average. Relative membrane conductivity (%) was calculated as follows:

$$\text{Relative membrane conductivity (\%)} = (\text{P1} - \text{P0})/(\text{P2} - \text{P0}) \times 100\,\% \quad (1)$$

where P0 represents the blank electrical conductivity (distilled water).

2.2.5. Total RNA Extraction, cDNA Synthesis, and Quantitative Real-Time PCR (q-PCR)

Total RNA was extracted from the button mushrooms using Omega's Plant RNA kit, albeit with minor alterations. The Nanodrop 2000 spectrophotometer (Thermo Scientific, Vacaville, CA, USA) was used for assessing RNA purity, followed by 1% (w/v) agarose gel electrophoresis to detect RNA degradation or contamination. The total RNA was then reverse transcribed into first-strand cDNA by using a HiScript II kit (Vazyme Biotech Co., Ltd., Nanjing, China).

Total RNA was collected and analyzed using qRT-PCR to determine the expression patterns of enzymatic browning and ethylene production genes in samples throughout storage. The four enzymatic browning-related genes and five ethylene synthesis-related genes in the button mushroom are *AbPPO1*, *AbPPO3*, *AbLAC10*, *AbLAC11*, *AbACO1*, *AbSAMS1*, *AbSAMS2*, *AbACS1*, and *AbACS2* and specific nucleotide sequence information was retrieved from the GenBank databases. The Light Cycler TM 96 System (Roche Molecular Systems, Inc., Mannheim, Germany) was utilized to conduct a quantitative real-time polymerase chain reaction. Table S1 (which is included in the Supplementary Materials) lists the primers used for EF1-α, the reference gene.

2.2.6. Data Processing

The experiments underwent three rounds of analysis, and results appeared as mean $\pm$ standard deviation. Researchers employed Microsoft Excel for statistical evaluations of all experimental data, calculating standard deviations accordingly. Plots were generated using Microsoft Excel, while IBM SPSS Statistics 27 software facilitated variance analysis across all article data, utilizing Tukey's HSD test for multiple comparisons. Statistical significance was determined at $p < 0.05$.

3. Results

3.1. Effects of CA on Visual Appearance and Quality Characteristics of *A. bisporus* During Storage

3.1.1. Effects of CA on Browning of *A. bisporus* During Storage

After harvesting, the external and internal color of button mushrooms in all CA groups and the control group gradually darkened with increasing storage time (Figures 1 and 2). Figure 1 showed that no significant browning occurred on the button mushroom peels of the five groups within the first 10 days of storage. Notably, browning symptoms initially emerged in both control and CA samples at day 11, with the extent of browning increasing with storage duration. From day 11 onwards until the experiment concluded, the coloration of samples stored at 1–3% O_2 and 15–17% CO_2 (CAII) and 1–3% O_2 and 13–15% CO_2 (CAI) groups remained notably brighter compared to the control group. At the conclusion of the storage period, button mushrooms under 1–3% O_2 and 15–17% CO_2 (CAII) browned and senesced more slowly than the control group. On day 12, exterior browning occurred in the following order: cap surface, CAII < CAI < CAIV < Control < CAIII.

External browning appeared early in button mushrooms stored in both CA and control groups, while internal browning was repressed (Figure 2). Flesh browning symptoms were evident in button mushrooms kept under a controlled atmosphere and in the control after 11 days of storage, with a notable difference from controls primarily emerging between days 12 and the end of the experiment. On day 13, 1–3% O_2 and 15–17% CO_2 (CAII) storage retained the best color, only slightly darker than fresh button mushrooms. Button mushrooms stored at 3–5% O_2 and 15–17% CO_2 (CAIV group) exhibited comparable color to those kept in ambient air.

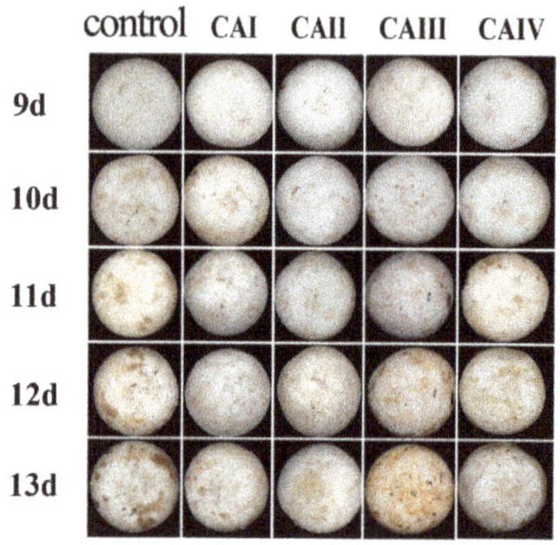

Figure 1. Effect of controlled atmosphere on the external appearance of *A. bisporus* during storage.

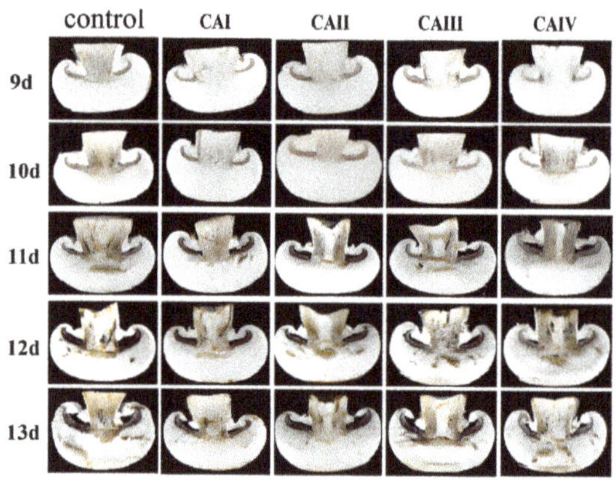

Figure 2. Effect of controlled atmosphere on the internal appearance of *A. bisporus* flesh during storage.

According to Figure 3a, the L* value of button mushroom caps stored in the five groups decreased over time, exhibiting rapid drops during the early and middle storage stages, with a slower reduction later. Button mushrooms maintained in a controlled atmosphere displayed a suppression of L* value reduction, particularly prominent in the middle and late phases. After storage, the L* values of button mushrooms in the 1–3% O_2 and 13–15% CO_2 (CAI), 1–3% O_2 and 15–17% CO_2 (CAII), and 3–5% O_2 and 15–17% CO_2 (CAIV) groups increased by 27.5%, 31.1%, and 22.4% compared to the button mushrooms in the control group, respectively. Button mushrooms in the 1–3% O_2 and 15–17% CO_2 (CAII) group performed better than either group, with a notable difference from controls primarily emerging between days 11 and the end of the experiment ($p < 0.05$). This result indicated that CA helped to postpone the external browning of button mushroom.

As shown in Figure 3b, the L* value of the flesh of button mushrooms stored in the five groups gradually decreased over time, with a rapid decline in the late stages. The difference between button mushrooms stored under a controlled atmosphere and controls reached statistical significance on day 10 of storage until the end of the experiment ($p < 0.05$). On day 11 and day 12, the 1–3% O_2 and 13–15% CO_2 (CAI) and 1–3% O_2 and 15–17%

CO_2 (CAII) groups significantly enhanced the L* value of the flesh of button mushrooms compared to the control, 3–5% O_2 and 13–15% CO_2 (CAIII), and 3–5% O_2 and 15–17% CO_2 (CAIV) groups. At the conclusion of storage, the L* value of the flesh of button mushrooms in the control group was 62, which was 21.1%, 20.9%, 8.1%, and 16.1% lower than that of the 1–3% O_2 and 13–15% CO_2 (CAI), 1–3% O_2 and 15–17% CO_2 (CAII), 3–5% O_2 and 13–15% CO_2 (CAIII), and 3–5% O_2 and 15–17% CO_2 (CAIV) groups, respectively. Together, these results provide important insights demonstrating that the CA successfully postponed the internal browning procedure.

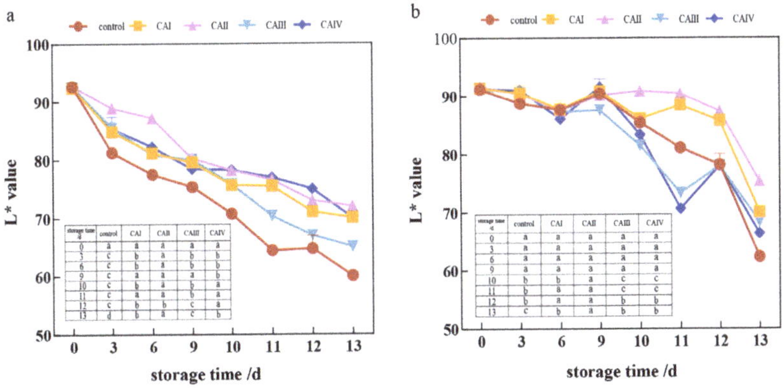

Figure 3. Effect of controlled atmosphere on the L* value of *A. bisporus* during storage: (**a**) surface and (**b**) flesh. The error bars represent the standard deviation (SD) for three replicates. In the table within the image, the lowercase letters a–d indicate statistically significant differences among the five groups ($p < 0.05$).

3.1.2. Effects of CA on Weight Loss and Hardness of *A. bisporus* During Storage

As demonstrated by Figure 4a, the weight loss rate of button mushrooms stored in the five groups increased over time, with a significant increase at the final stage. On day 13 of storage, the weight loss rate of control button mushrooms rose by 34.6%, 32.7%, and 38.4% compared to button mushrooms stored in the 1–3% O_2 and 13–15% CO_2 (CAI), 1–3% O_2 and 15–17% CO_2 (CAII), and 3–5% O_2 and 15–17% CO_2 (CAIV) groups, respectively. The 3–5% O_2 and 15–17% CO_2 (CAIV) storage outperformed other groups with a significant difference observed from day 11 until the end of the investigation ($p < 0.05$). Specifically, at approximately 11–13 days of storage, the weight loss rate of button mushrooms under 3–5% O_2 and 15–17% CO_2 (CAIV) increased from 5.8% to 7.8%. However, the weight loss rate of samples in 3–5% O_2 and 13–15% CO_2 (CAIII) group increased significantly from day 10 of storage until the completion of the study, surpassing the other four groups ($p < 0.05$). This was most likely due to the low CO_2 situation, which increased the dry-matter consumption of button mushrooms after harvesting, perhaps increasing the weight loss rate. The data presented here indicate that properly controlled atmosphere storage significantly reduces weight loss in button mushrooms.

As illustrated in Figure 4b, the hardness of the button mushroom stored in the five groups decreased over time, showing a sharp decline in the early and late storage phases, with a gradual decrease in the middle phase. On day 9, button mushrooms in the control group had a hardness of 17.6 N, which was 22.8%, 21.2%, 1.2%, and 1.1% lower than the 1–3% O_2 and 13–15% CO_2 (CAI), 1–3% O_2 and 15–17% CO_2 (CAII), 3–5% O_2 and 13–15% CO_2 (CAIII), and 3–5% O_2 and 15–17% CO_2 (CAIV) groups, respectively. At the conclusion of storage, the hardness of button mushrooms under 1–3% O_2 and 13–15% CO_2 (CAI), 1–3% O_2 and 15–17% CO_2 (CAII), 3–5% O_2 and 13–15% CO_2 (CAIII), and 3–5% O_2 and 15–17% CO_2 (CAIV) was 14.9 N, 14.2 N, 9.6 N, and 12.8 N, respectively, while the hardness of the control button mushrooms was 12.1 N. Higher carbon dioxide levels in the 1–3% O_2 and 15–17% CO_2 (CAII) group may reduce respiratory activity and water loss in button mushrooms, causing softening to take longer. This finding provides vital insights into higher CO_2 concentrations proving useful in maintaining the hardness of button mushrooms during storage.

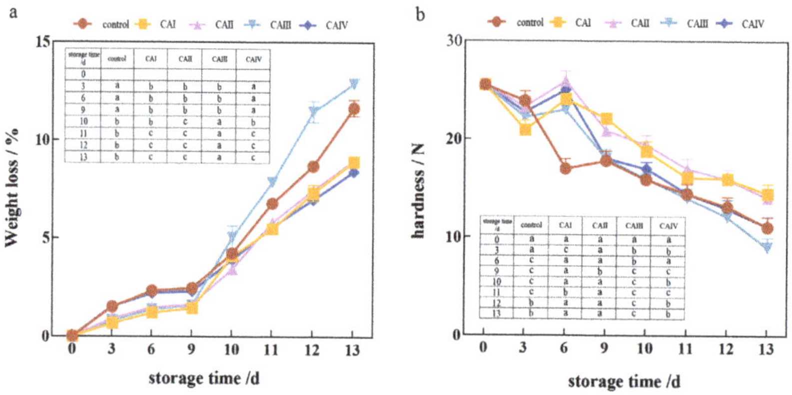

Figure 4. Effect of controlled atmosphere on the weight loss and hardness of *A. bisporus* during storage: (**a**) weight loss and (**b**) hardness. The error bars represent the standard deviation (SD) for three replicates. In the table within the image, the lowercase letters a–c indicate statistically significant differences among the five groups ($p < 0.05$).

3.1.3. Effects of CA on Cell Membrane Permeability of *A. bisporus* During Storage

Figure 5 shows that the cell membrane permeability of the control group increased significantly with increasing storage duration, particularly after the ninth day, which could be attributed to the browning process. The permeability of cell membranes in samples stored under a controlled atmosphere was significantly lower than that of control samples, with a statistically significant difference observed from day 6 to the end of the experiment ($p < 0.05$). Specifically, control button mushrooms had 1.3, 1.4, and 1.1 times increased cell membrane permeability at the end of storage compared to button mushrooms storage under 1–3% O_2 and 13–15% CO_2 (CAI), 1–3% O_2 and 15–17% CO_2 (CAII), and 3–5% O_2 and 15–17% CO_2 (CAIV), respectively. This suggests that controlled atmosphere storage effectively preserves the integrity of the cell membrane, reducing the browning of tissues [29].

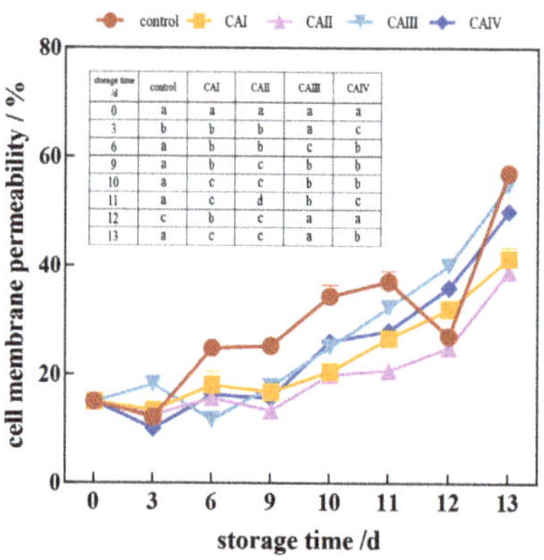

Figure 5. Effect of controlled atmosphere on the cell membrane permeability of *A. bisporus* during storage. The error bars represent the standard deviation (SD) for three replicates. In the table within the image, the lowercase letters a–d indicate statistically significant differences among the five groups ($p < 0.05$).

3.2. Expression Patterns of Enzymatic Browning-Related Genes

The mRNA levels of two polyphenol oxidase (PPO) and two lactase (LAC) genes in button mushrooms stored in the 1–3% O_2 and 13–15% CO_2 (CAI), 1–3% O_2 and 15–17% CO_2 (CAII), 3–5% O_2 and 13–15% CO_2 (CAIII), and 3–5% O_2 and 15–17% CO_2 (CAIV) groups for various times (0, 3, 6, 9, 10, 11, 12, and 13 days) (Figure 6) were investigated.

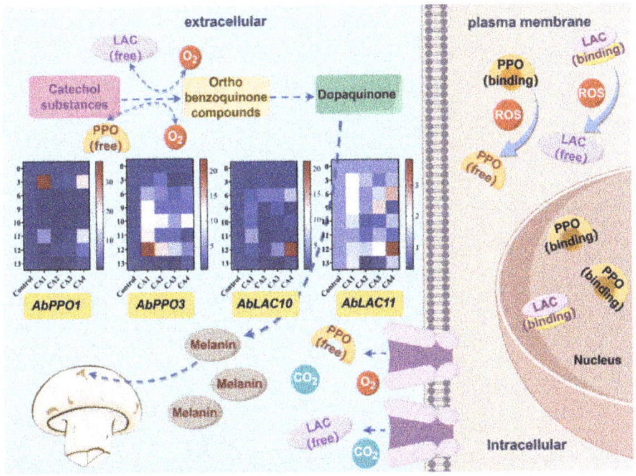

Figure 6. Effect of controlled atmosphere on the expression patterns of enzymatic browning-related genes of *A. bisporus* during storage (by Figdraw).

The 1–3% O_2 and 15–17% CO_2 (CAII) conditions reduced *AbPPO1* gene transcription in button mushrooms compared to the control (Figure 6). Significant differences were observed from day 6 onwards until the experiment concluded ($p < 0.05$). The 1–3% O_2 and 13–15% CO_2 (CAI) and 3–5% O_2 and 13–15% CO_2 (CAIII) storage significantly increased *AbPPO1* gene transcription in the first 3 days of storage compared to the control group ($p < 0.05$). This was most likely due to extrinsic impacts such as the removal of roots after harvesting, which might induce the expression of the *AbPPO1* gene [30]. Furthermore, after 6–13 days of storage, the transcription level of the *AbPPO1* gene reduced considerably in the 3–5% O_2 and 13–15% CO_2 (CAIII) group when compared to the control group ($p < 0.05$). Thus, the finding shows that *AbPPO1* was involved in the browning of the button mushroom.

The transcription levels of the *AbPPO3* gene were markedly higher in all four controlled atmosphere groups from day 6 to day 12 of storage ($p < 0.05$), which corresponded to the visual appearance (Figures 1 and 2) and browning degree (Figure 3a,b). On day 13 of storage, 1–3% O_2 and 15–17% CO_2 (CAII) and 3–5% O_2 and 13–15% CO_2 (CAIII) significantly decreased *AbPPO3* transcription levels, indicating that 1–3% O_2 and 15–17% CO_2 (CAII) and 3–5% O_2 and 13–15% CO_2 (CAIII) may reduce *AbPPO3* activity in later storage stages.

The transcription patterns of the *AbLAC10* and *AbLAC11* genes are depicted in Figure 6. Throughout the storage period, the transcription levels of the *AbLAC10* and *AbLAC11* genes in all four CA groups increased initially, decreased, and then increased again. The transcription levels of *AbLAC11* and *AbLAC10* genes in button mushrooms were significantly lower under 1–3% O_2 and 15–17% CO_2 (CAII) when compared to the control from day 6 onwards until the experiment concluded ($p < 0.05$). Moreover, over 10–13 days of storage, the expression of *AbLAC10* and *AbLAC11* genes in CAII-stored button mushrooms increased from 0.6 to 0.9 and 0.6 to 1.4, respectively. Notably, 1–3% O_2 and 15–17% CO_2 (CAII) was the most effective in diminishing gene expressions of *AbLAC10* and *AbLAC11* after day 6 of storage.

3.3. Expression Patterns of Ethylene Synthesis-Related Genes

The expression patterns of ethylene synthesis-related genes (*AbACO1, AbSAMS1, AbSAMS2, AbACS1*, and *AbACS2*) in button mushrooms stored in 1–3% O_2 and 13–15% CO_2 (CAI), 1–3% O_2 and 15–17% CO_2 (CAII), 3–5% O_2 and 13–15% CO_2 (CAIII), and 3–5% O_2 and 15–17% CO_2 (CAIV) groups for various times (0, 3, 6, 9, 10, 11, 12, and 13 days) were studied (Figure 7).

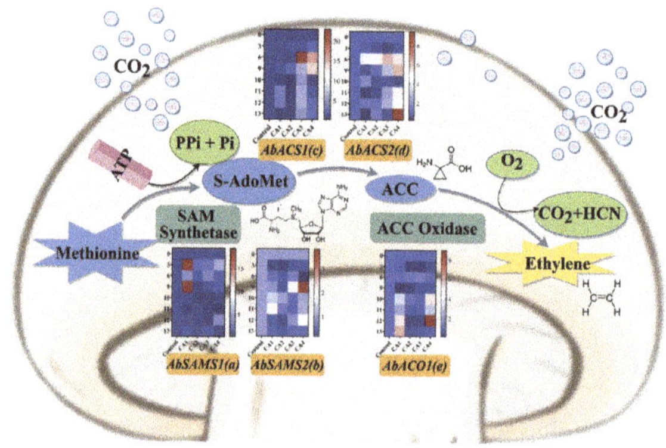

Figure 7. Effect of controlled atmosphere on the expression patterns of ethylene synthesis-related genes of *A. bisporus* during storage (by Figdraw): (**a**) *AbSAMS1*, (**b**) *AbSAMS2*, (**c**) *AbACS1*, (**d**) *AbACS2*, and (**e**) *AbACO1*.

Figure 7a shows that from day 3 to day 9 of storage, the transcription level of the *AbSAMS1* gene in the 1–3% O_2 and 13–15% CO_2 (CAI) storage was significantly higher than that in the control group ($p < 0.05$), suggesting that this storage method may have an upregulation effect on the expression of this gene. In contrast, the expression levels under other storage conditions, such as in the 1–3% O_2 and 15–17% CO_2 (CAII) and C3–5% O_2 and 13–15% CO_2 (CAIII) groups, displayed distinct patterns throughout the storage period. The *AbSAMS1* gene expression under 1–3% O_2 and 15–17% CO_2 (CAII) and 3–5% O_2 and 13–15% CO_2 (CAIII) increased from 0.7 to 0.9, and 0.9 to 1.5 from day 6 onwards until the experiment concluded, respectively.

The transcription level of the *AbSAMS2* gene exhibited significant variations in samples stored in all four groups during storage at 4 °C (Figure 7b), peaking in the middle stages and declining towards the end. The expression level in the 3–5% O_2 and 15–17% CO_2 (CAIV) group surged dramatically on day 9 of storage, surpassing the other groups ($p < 0.05$). Conversely, after 10 to 13 days of storage, the transcription level of the *AbSAMS2* gene showed a marked decline in the 1–3% O_2 and 15–17% CO_2 (CAII) group relative to the control samples ($p < 0.05$). Furthermore, the observed reduction in transcription levels suggests that 1–3% O_2 and 15–17% CO_2 (CAII) can inhibit the expression of the *AbSAMS2* gene.

The transcription patterns of *AbACS1* and *AbACS2* genes were analogous in samples stored in all four groups (Figure 7c,d). The transcription of the *AbACS1* and *AbACS2* genes was significantly inhibited in the button mushrooms stored in the 1–3% O_2 and 13–15% CO_2 (CAI) and 1–3% O_2 and 15–17% CO_2 (CAII) groups ($p < 0.05$), which corresponds to the visual appearance. However, the transcription levels of the two genes increased until nine days in button mushrooms under 3–5% O_2 and 13–15% CO_2 (CAIII) and 3–5% O_2 and 15–17% CO_2 (CAIV) storage ($p < 0.05$). This was probably due to a deeper senescence of button mushrooms during later storage, which increased ethylene production and could trigger the expression of *AbACS1* and *AbACS2* genes. After 10–13 days of storage, CAII-stored button mushrooms showed an increase in *AbACS1* and *AbACS2* gene expression from 0.25 to 0.65 and 0.7 to 1.4, respectively.

The *AbACO1* gene transcription level was lower in 1–3% O_2 and 15–17% CO_2 (CAII) and 3–5% O_2 and 13–15% CO_2 (CAIII) groups compared to the control group (Figure 7e), showing a statistically significant difference from day 6 of storage through to the end of the trial ($p < 0.05$). Conversely, the transcription level of *AbACO1* was significantly heightened in 1–3% O_2 and 13–15% CO_2 (CAI) and 3–5% O_2 and 15–17% CO_2 (CAIV) groups during the later storage stages ($p < 0.05$). This suggests that the transcription of the *AbACO1* gene could be induced by external ethylene in the environment. Moreover, button mushrooms stored in the 1–3% O_2 and 15–17% CO_2 (CAII) group could inhibit the expression of *AbACO1*, and the inhibitory effect was particularly pronounced in the middle stage, aligning with visual assessments.

4. Discussion

Controlled atmosphere storage extends the storage period of fruits and vegetables by increasing the concentration of carbon dioxide (CO_2) and/or decreasing the concentration of oxygen (O_2) in the storage room [31]. Compared to horticultural products, edible fungi have some unique characteristics, such as high postharvest respiration and resilience to external carbon dioxide levels. Previous studies indicated that, unlike the low carbon dioxide atmospheres typically used for fruits and vegetables, high carbon dioxide levels may help slow down the senescence process of button mushrooms and extend their postharvest shelf life [18]. Furthermore, a high carbon dioxide environment impedes ethylene production and can also regulate the respiration rate and energy metabolism of button mushrooms. These conditions can synergize with low oxygen levels to manage physiological changes during the postharvest storage or packaging of various fresh products [32,33]. In the present study, observing the visual traits of button mushrooms in storage reveals that controlled atmosphere storage can reduce L* value decrease while maintaining button mushroom appearance, thus ensuring an appealing presentation for sale. Gene expression analysis indicated that 1–3% O_2 and 15–17% CO_2 (CAII) storage, characterized by low oxygen and high carbon dioxide levels, effectively mitigates browning and decreases the transcription levels of enzymatic browning-related genes (*AbPPO1*, *AbPPO3*, *AbLAC10*, and *AbLAC11*, respectively) in the samples.

Browning in stored button mushrooms involves an intricate chemical process, in which phenolic compounds undergo enzymatic oxidation to form quinones, which subsequently react to produce melanin [7]. This process may stem from compromised membrane integrity, as indicated by increased electrolyte leakage and malondialdehyde (MDA) accumulation [34]. Specifically, the permeability of cell membranes in samples was found to be 1.3, 1.5, and 1.1 times higher in the control than in button mushrooms stored in 1–3% O_2 and 13–15% CO_2 (CAI), 1–3% O_2 and 15–17% CO_2 (CAII), and 3–5% O_2 and 15–17% CO_2 (CAIV) groups, respectively, at the end of storage. This observation aligns with findings by Ayala, A [35]. Additionally, the heightened expression of the *PPO* gene and related enzyme activity may drive the oxidation of phenols, contributing to the browning of button mushrooms after harvest during storage. Furthermore, Lin et al. demonstrated that elevated CO_2 levels inhibited browning in button mushrooms by preserving cellular integrity and inhibiting phenolic-related enzymes, such as PPO [36]. This finding aligns with the current research, showing reduced cell membrane permeability and *PPO* expression in high CO_2 environments. Additionally, it is crucial to comprehend how environmental factors interact with gene expression to clarify the pathways involved in the oxidative stress response. Future research should focus on the impact of various gas conditions on the expression of *PPO* and *LAC* genes using both in vitro and in vivo models.

Senescence substantially affects the storage duration of button mushroom. Evidence indicates that ethylene release plays a crucial role in the maturation and senescence of bulk horticultural products, with ethylene synthesis also speeding up these processes in edible fungi. As such, understanding the mechanisms of ethylene production and its impact on button mushrooms is vital to developing appropriate storage techniques. Ethylene biosynthesis occurs when the presence of ethylene, either from internal or external sources, boosts the expression of *ACO* and *ACS* genes during the ripening of climacteric fruits [37]. Studies indicate that ethylene production in the button mushroom aligns with the ACC pathway of plants [38]. The primary mechanism involves the initial transformation of methionine into S-adenosylmethionine (SAM), facilitated by adenosine triphosphate (ATP) and SAM synthase (SAMS). Subsequently, SAM undergoes conversion to 1-aminocyclopropane-1-carboxylic acid (ACC) through the action of ACC synthase (ACS). Subsequently, ACC converts to ethylene facilitated by ACC oxidase (ACO). ACC synthase (ACS) and ACC oxidase (ACO) are the key enzymes in ethylene biosynthesis. Ethylene production correlates positively with the transcription products of the genes linked to ethylene synthesis enzymes (ACS or ACO) [39]. In tomato, the *ACS* genes *SlACS2* and *SlACS4* and *ACO* genes *SlACO1* and *SlACO4* play significant roles in ethylene synthesis [40]. Additionally, the promoters of *LeACS2* and *LeACS4* carry response elements for ethylene, wounding, and anaerobic conditions [41]. In the current work, we aimed to uncover the molecular mechanisms through which ethylene influences the postharvest maturation and senescence of button mushrooms. Based on software predictions, we identified ethylene response elements in the genes *AbACO1*, *AbSAMS1*, *AbSAMS2*, *AbACS1*, and *AbACS2* [25]. Compared to the control group, controlled atmosphere (CA) storage inhibited the expression of the *SAMS*, *ACO*, and *ACS* genes. Similar results have been reported in tomatoes [40]. Notably, 1–3% O_2 and 15–17% CO_2 (CAII) significantly reduced the expression of ethylene synthesis-related enzymes after 6 days of storage. The findings suggest that ethylene plays a regulatory role in the maturation and senescence of postharvest button mushrooms, resonating with climacteric fruit behavior. Furthermore, this suggests that altering the atmospheric composition could potentially influence ethylene production, subsequently affecting button mushroom growth patterns and physiological responses.

To conclude, high CO_2 controlled atmosphere storage preserved the visual quality of button mushrooms, while inhibiting weight loss and softening. Using 1–3% O_2 and 15–17% CO_2 (CAII) suppressed the transcription levels of *PPO* and *LAC*, as well as ethylene synthesis-related genes, thus improving the browning and senescence of button mushrooms. These findings indicate that high CO_2 controlled atmosphere storage has the potential to improve the nutrition, quality, and ultimately consumer preference for button mushrooms after harvesting and during storage. Additionally, tailored CA conditions for button mushrooms could be further optimized to improve their shelf life and overall quality.

5. Conclusions

Based on observations regarding the weight loss, L* value, hardness, cell membrane permeability, and visual appearance, CA storage maintained the storage quality of the button mushrooms over extended periods. In comparison to the other three CA groups, 1–3% O_2 and 13–15% CO_2 (CAI), 1–3% O_2 and 15–17% CO_2 (CAII), and 3–5% O_2 and 15–17% CO_2 (CAIV), the 1–3% O_2 and 15–17% CO_2 (CAII) group demonstrated the most favorable outcomes. In the meantime, the expression patterns of the enzyme-catalyzed browning-related genes *AbPPO1*, *AbLAC10*, and *AbLAC11* matched the visual appearance and browning degree, which were effectively inhibited under a high concentration of CO_2. Additionally, results indicated that 1–3% O_2 and 15–17% CO_2 (CAII) lowered the expression of SAMS, ACO, and ACS enzyme-associated genes, potentially postponing the senescence of postharvest button mushrooms. What emerges from these results reported here is that high CO_2 controlled atmosphere storage can delay button mushroom browning and ethylene synthesis, consequently increasing the postharvest shelf life of button mushrooms. Further investigations are warranted to explore the mechanisms underlying these effects, specifically focusing on the metabolic pathways influenced by elevated carbon dioxide concentrations.

Supplementary Materials: The following supporting information can be downloaded at: https://www.mdpi.com/article/10.3390/foods13213486/s1, Table S1: Primer names and sequence.

Author Contributions: Y.Y., conceptualization, methodology, software, validation, formal analysis, writing; Y.L., investigation, formal analysis; O.J., conceptualization, supervision, validation; B.F., conceptualization, supervision, validation; M.C., supervision, validation, funding acquisition; J.M., supervision, validation. B.D., conceptualization, supervision, visualization, funding acquisition. All authors have read and agreed to the published version of the manuscript.

Funding: This research was supported by the following projects: This study was funded by the National Science Foundation of China (No.32102053); the Distinguished and Excellent Young Scholar Cultivation Project of Shanxi Agricultural University (2022YQPYGC08); the Key Scientific and Technological Innovation Team of Edible Fungi of Shanxi Province (201805D131009); and the Major Special Science and Technology Projects of Shanxi Province (202301140601015).

Institutional Review Board Statement: Not applicable.

Informed Consent Statement: Not applicable.

Data Availability Statement: The original contributions presented in the study are included in the article/Supplementary Material, further inquiries can be directed to the corresponding authors.

Conflicts of Interest: The authors declare no conflicts of interest.

References

1. Wu, Y.; Hu, Q.; Li, Z.; Pei, F.; Mariga, A.; Yang, W. Effect of nanocomposite-based packaging on microstructure and energy metabolism of *Agaricus bisporus*. *Food Chem.* **2019**, *276*, 790–796. [CrossRef] [PubMed]
2. Yang, X.; Zhang, Y.; Kong, Y.; Zhao, J.; Sun, Y.; Huang, M. Comparative analysis of taste compounds in shiitake mushrooms processed by hot-air drying and freeze drying. *Int. J. Food Prop.* **2019**, *22*, 1100–1111. [CrossRef]
3. Mohapatra, D.; Bira, Z.M.; Kerry, J.P.; Frías, J.M.; Rodrigues, F.A. Postharvest hardness and color evolution of white button mushrooms (*Agaricus bisporus*). *J. Food Sci.* **2010**, *75*, E146–E152. [CrossRef]
4. Braaksma, A.; Schaap, D.J.; Donkers, J.W.; Schipper, C.M.A. Effect of cytokinin on cap opening in *Agaricus bisporus* during storage. *Postharvest Biol. Technol.* **2001**, *23*, 171–173. [CrossRef]
5. Castellanos-Reyes, K.; Villalobos-Carvajal, R.; Beldarrain-Iznaga, T. Fresh mushroom preservation techniques. *Foods* **2021**, *10*, 2126. [CrossRef]
6. Sharma, M.; Suman, B.; Gupta, D. Development of *Agaricus bisporus* hybrids and their evaluation for higher yield. *Indian J. Hortic.* **2016**, *73*, 550–556. [CrossRef]
7. Wu, X.; Guan, W.; Yan, R.; Lei, J.; Xu, L.; Wang, Z. Effects of UV-C on antioxidant activity, total phenolics and main phenolic compounds of the melanin biosynthesis pathway in different tissues of button mushroom. *Postharvest Biol. Technol.* **2016**, *118*, 51–58. [CrossRef]
8. Li, N.; Chen, F.; Cui, F.; Sun, W.; Zhang, J.; Qian, L.; Yang, Y.; Wu, D.; Dong, Y.; Jiang, J.; et al. Improved postharvest quality and respiratory activity of straw mushroom (*Volvariella volvacea*) with ultrasound treatment and controlled relative humidity. *Sci. Hortic.* **2017**, *225*, 56–64. [CrossRef]
9. Gao, M.; Feng, L.; Jiang, T. Browning inhibition and quality preservation of button mushroom (*Agaricus bisporus*) by essential oils fumigation treatment. *Food Chem.* **2014**, *149*, 107–113. [CrossRef]
10. Dokhanieh, A.Y.; Aghdam, M.S. Postharvest browning alleviation of *Agaricus bisporus* using salicylic acid treatment. *Sci. Hortic.* **2016**, *207*, 146–151. [CrossRef]
11. Ozturka, B.; Havsuta, E.; Yildizb, K. Delaying the postharvest quality modifications of *Cantharellus cibarius* mushroom by applying citric acid and modified atmosphere packaging. *LWT* **2021**, *138*, 110639. [CrossRef]
12. Meng, D.; Zhang, Y.; Yang, R.; Wang, J.; Zhang, X.-H.; Sheng, J.; Wang, J.; Fan, Z. Arginase participates in the methyl jasmonate-regulated quality maintenance of postharvest *Agaricus bisporus* fruit bodies. *Postharvest Biol. Technol.* **2017**, *132*, 7–14. [CrossRef]
13. Bodbodak, S.; Moshfeghifar, M. Advances in controlled atmosphere storage of fruits and vegetables. In *Eco-Friendly Technology for Postharvest Produce Quality*; Academic Press: Cambridge, MA, USA, 2016; pp. 39–76.
14. Wang, J.; Yan, P.; MA, H.; Feng, W.; Wang, D. Effect of modified atmosphere package conditions on preservation of green walnut fruit and kernel traits. *Mod. Food Sci. Technol.* **2014**, *30*, 169–176.
15. Yi, J.; Feng, H.; Bi, J.; Zhou, L.; Zhou, M.; Cao, J.; Li, J. High hydrostatic pressure induced physiological changes and physical damages in asparagus spears. *Postharvest Biol. Technol.* **2016**, *118*, 1–10. [CrossRef]
16. Belay, Z.A.; Caleb, O.J.; Opara, U.L. Impacts of low and super-atmospheric oxygen concentrations on quality attributes, phytonutrient content and volatile compounds of minimally processed pomegranate arils (cv. wonderful). *Postharvest Biol. Technol.* **2017**, *124*, 119–127. [CrossRef]
17. Li, L.; Sun, H.; Kitazawa, H.; Wang, X. Effects of a high O_2 dynamic-controlled atmosphere technology on the browning of postharvest white mushroom (*Agaricus bisporus*) in relation to energy metabolism. *Food Sci. Technol. Int.* **2017**, *23*, 385–395. [CrossRef]

18. Amodio, M.; Colelli, G. Controlled-atmosphere storage of fresh-cut 'cardoncello' mushrooms (*Pleurotus eryngii*). *Acta Hortic.* **2003**, *599*, 731–735. [CrossRef]
19. Navina, B.; Huthaash, K.K.; Velmurugan, N.K.; Korumilli, T. Insights into recent innovations in anti-browning strategies for fruit and vegetable preservation. *Trends Food Sci. Technol.* **2023**, *139*, 104128. [CrossRef]
20. Iqbal, A.; Murtaza, A.; Hu, W.; Ahmad, I.; Ahmed, A.; Xu, X. Activation and inactivation mechanisms of polyphenol oxidase during thermal and non-thermal methods of food processing. *Food Bioprod. Process.* **2019**, *117*, 170–182. [CrossRef]
21. Vámos-Vigyázó, L. Polyphenol oxidase and peroxidase in fruits and vegetables. *Crit. Rev. Food Sci. Nutr.* **1981**, *15*, 49–127. [CrossRef]
22. Nicolas, J.J.; Richard-Forget, F.C.; Goupy, P.M.; Amiot, M.-J.; Aubert, S.Y. Enzymatic browning reactions in apple and apple products. *Crit. Rev. Food Sci. Nutr.* **1994**, *34*, 109–157. [CrossRef] [PubMed]
23. Marçal, S.; Sousa, A.S.; Taofiq, O.; Antunes, F.; Morais, A.M.M.B.; Freitas, A.C.; Barros, L.; Ferreira, I.C.F.R.; Pintado, M. Impact of postharvest preservation methods on nutritional value and bioactive properties of mushrooms. *Trends Food Sci. Technol.* **2021**, *110*, 418–431. [CrossRef]
24. Wheeler, R.M.; Peterson, B.V.; Stutte, G.W. Ethylene production throughout growth and development of plants. *HortScience* **2004**, *39*, 1541–1545. [CrossRef]
25. Li, T.; Zhang, J.; Gao, X.; Chen, J.; Qiu, L. The molecular mechanism for the ethylene regulation of postharvest button mushrooms maturation and senescence. *Postharvest Biol. Technol.* **2019**, *156*, 110930. [CrossRef]
26. Sun, Y.; Li, W. Effects of different cling films on freshness of *Pleurotus ostreatus*. *J. Food Meas. Charact.* **2017**, *11*, 592–597. [CrossRef]
27. Mu, H.; Gao, H.; Chen, H.; Fang, X.; Han, Q. A novel controlled release ethanol emitter: Preparation and effect on some postharvest quality parameters of Chinese bayberry during storage. *J. Sci. Food Agric.* **2017**, *97*, 4929–4936. [CrossRef]
28. Wantat, A.; Seraypheap, K.; Rojsitthisak, P. Effect of chitosan coatings supplemented with chitosan-montmorillonite nanocomposites on postharvest quality of 'Hom Thong' banana fruit. *Food Chem.* **2022**, *374*, 131731. [CrossRef]
29. Liu, Z.; Wang, X. Changes in color, antioxidant, and free radical scavenging enzyme activity of mushrooms under high oxygen modified atmospheres. *Postharvest Biol. Technol.* **2012**, *69*, 1–6. [CrossRef]
30. Li, N.-y.; Cai, W.-m.; Jin, Q.-l.; Qin, Q.-p.; Ran, F.-l. Molecular cloning and expression of polyphenol oxidase genes from the mushroom, *Agaricus bisporus*. *Agric. Sci. China* **2011**, *10*, 185–194. [CrossRef]
31. Sharples, R.O. Book Reviews: Postharvest Physiology of Perishable Plant Products. *Outlook Agric.* **1992**, *21*, 315–316. [CrossRef]
32. Blanch, M.; Rosales, R.; Palma, F.; Sanchez-Ballesta, M.T.; Escribano, M.I.; Merodio, C. CO_2-driven changes in energy and fermentative metabolism in harvested strawberries. *Postharvest Biol. Technol.* **2015**, *110*, 33–39. [CrossRef]
33. Lumpkin, C.; Fellman, J.K.; Rudell, D.R.; Mattheis, J.P. 'Fuji' apple (*Malus domestica* Borkh.) volatile production during high CO_2 controlled atmosphere storage. *Postharvest Biol. Technol.* **2015**, *100*, 234–243. [CrossRef]
34. Shekari, A.; Hassani, R.N.; Aghdam, M.S.; Rezaee, M.; Jannatizadeh, A. The effects of melatonin treatment on cap browning and biochemical attributes of *Agaricus bisporus* during low temperature storage. *Food Chem.* **2021**, *348*, 129074. [CrossRef]
35. Ayala, A.; Muñoz, M.F.; Argüelles, S. Lipid peroxidation: Production, metabolism, and signaling mechanisms of malondialdehyde and 4-hydroxy-2-nonenal. *Oxidative Med. Cell. Longev.* **2014**, *2014*, 360438. [CrossRef] [PubMed]
36. Lin, Q.; Lu, Y.; Zhang, J.; Liu, W.; Guan, W.; Wang, Z. Effects of high CO_2 in-package treatment on flavor, quality and antioxidant activity of button mushroom (*Agaricus bisporus*) during postharvest storage. *Postharvest Biol. Technol.* **2017**, *123*, 112–118. [CrossRef]
37. Pattyn, J.; Vaughan-Hirsch, J.; Van de Poel, B. The regulation of ethylene biosynthesis: A complex multilevel control circuitry. *New Phytol.* **2021**, *229*, 770–782. [CrossRef]
38. Zhang, C.; Huang, T.; Shen, C.; Wang, X.; Qi, Y.; Shen, J.; Song, A.; Qiu, L.; Ai, Y. Downregulation of ethylene production increases mycelial growth and primordia formation in the button culinary-medicinal mushroom, *Agaricus bisporus* (Agaricomycetes). *Int. J. Med. Mushrooms* **2016**, *18*, 1131–1140. [CrossRef]
39. Zhang, M.; Smith, J.A.C.; Harberd, N.P.; Jiang, C. The regulatory roles of ethylene and reactive oxygen species (ROS) in plant salt stress responses. *Plant Mol. Biol.* **2016**, *91*, 651–659. [CrossRef]
40. Yu, W.; Ma, P.; Sheng, J.; Shen, L. Arginine and cysteine delay postharvest ripening of tomato fruit by regulating ethylene production. *Postharvest Biol. Technol.* **2024**, *216*, 113052. [CrossRef]
41. Lincoln, J.E.; Campbell, A.D.; Oetiker, J.; Rottmann, W.H.; Oeller, P.W.; Shen, N.F. Le-acs4, a fruit ripening and wound-induced 1-aminocyclopropane-1-carboxylate synthase gene of tomato (*lycopersicon esculentum*). expression in escherichia coli, structural characterization, expression characteristics, and phylogenetic analysis. *J. Biol. Chem.* **1993**, *268*, 19422–19430. [CrossRef]

foods

MDPI

Article

Carvacrol Effectively Inhibits *Pseudomonas tolaasii* In Vitro and Induces Resistance to Brown Blotch Disease in Postharvest *Agaricus bisporus*

Lei Zhang [1,†], Rui Song [1,†], Zixuan Shi [1], Shuai Yuan [1], Lu Jiao [1], Mengsha Ma [1], Xing Wang [1], Lin Chen [2], Xia Liu [1] and Demei Meng [1,*]

1 Tianjin Key Laboratory of Food Quality and Health, College of Food Science and Engineering, Tianjin University of Science & Technology, Tianjin 300457, China; 22844990@mail.tust.edu.cn (L.Z.); 18331063572@163.com (R.S.); 22845913@mail.tust.edu.cn (Z.S.); yuanshuai@mail.tust.edu.cn (S.Y.); 22844934@mail.tust.edu.cn (L.J.); 23844962@mail.tust.edu.cn (M.M.); ax811@mail.tust.edu.cn (X.W.); liuxia@tust.edu.cn (X.L.)

2 School of Chemistry, Chemical Engineering and Biotechnology, Nanyang Technological University, Singapore 637459, Singapore; chen.lin@ntu.edu.sg

* Correspondence: mengdm@tust.edu.cn; Tel./Fax: +86-022-60912419

† These authors contributed equally to this work.

Abstract: Carvacrol (CAR), a naturally occurring phenolic monoterpene compound, has recently received attention for its potential use in food preservation. However, whether it is effective in controlling brown blotch disease caused by *Pseudomonas tolaasii* in edible mushrooms is unknown. The results of this study showed that CAR effectively inhibits and kills *P. tolaasii* in vitro by disrupting cell membrane integrity and causing the leakage of cellular components. Intracellular proteins and the DNA of *P. tolaasii* may not be the targets of CAR. CAR fumigation at a concentration as low as 20 μmol L^{-1} CAR effectively inhibited *P. tolaasii*-caused brown blotch disease in *Agaricus bisporus*, accompanied by a decrease in polyphenol oxidase activation, melanin production, and malondialdehyde accumulation. CAR treatment also significantly increased the activities of β-1,4-N-acetyl-glucosaminnidase, three antioxidant enzymes, and phenylpropanoid pathway-related enzymes, as well as promoting the accumulation of phenolic, flavonoid, and lignin substances in mushrooms, thereby inducing the resistance of mushrooms to the disease. These results demonstrate the potential application of carvacrol to control bacterial disease in *A. bisporus* mushrooms.

Keywords: antibacterial activity; action mechanism; brown blotch disease; carvacrol; disease resistance; edible mushrooms

1. Introduction

Agaricus bisporus (*A. bisporus*), also known as the white button mushroom, is one of the most widely cultivated mushroom species in the world, and its production accounts for about 1/6 of the total global mushroom production [1,2]. It has a huge market potential and research prospects by virtue of its rich nutritional value and increasing sales volume [3–5]. *A. bisporus* is susceptible to infection by *Pseudomonas*, which causes brown blotch disease during cultivation and postharvest transport and storage. Infected mushrooms not only show changes in appearance, flavor, and nutritional value, but their yield and shelf life are also negatively affected. In particular, *Pseudomonas tolasi* (*P. tolaasii*), a common soil bacterium that secretes an extracellular toxin called tolaasin, can cause brown, small, and irregular lesions on mushroom caps that may even coalesce to cover the entire surface of the mushroom [3,6]. Currently, the disease control method in mushrooms mainly focuses on low-temperature and chemical preservation (e.g., sodium hypochlorite, sodium sulfite, and chlorine dioxide), but there are limitations to its practical application in terms of ineffectiveness and safety risks [7]. Hence, there is a need to search for new alternative methods to reduce the development of this disease.

Single or multiple compounds derived from plant essential oils (EOs) are considered promising candidates for reducing the use of chemical agents due to their natural and potential antimicrobial properties [8,9]. Carvacrol (CAR), a naturally occurring phenolic monoterpene compound, is a major active ingredient in the OEs of plants such as thyme (*Thymus vulgaris*) and oregano (*Origanum vulgare*). CAR has been internationally certified for its safety and good antimicrobial activity [10]; it also has the advantages of favorable biocompatibility, a pleasant odor, lipophilic chemicals, and antioxidant properties [11–13]. CAR not only exerts inhibitory effects on a wide range of foodborne bacterial pathogens such as *Staphylococcus aureus*, *Escherichia coli*, and *Salmonella* [14,15] but also shows potent activity against many kinds of fungal pathogens, such as *Candida albicans* [16], *Valsa pyri* [17], *Penicillium citrinum* and *Meyerozyma caribbica* [18]. However, previous research on the efficacy of CAR against bacteria as a food preservative, mainly in the systems of meat and meat products [19,20], dairy products [21], and seafood [22], and its effectiveness against *P. tolaasii* in vitro and in harvested horticultural crops, especially postharvest mushrooms, remains unknown. In addition, CAR has great potential for controlling phytopathogenic diseases, such as pear plants [17] and fungal decay in postharvest agricultural products, such as red grapefruit [23], goji fruit [24], wheat [25], and tomato fruit [13]. However, compared to fruits and vegetables, there are relatively few reports on the preservation effect of CAR in edible mushrooms. In particular, edible mushrooms comprise different components to fruits and vegetables, and it is, therefore, uncertain whether the CAR treatment is effective against *P. tolaasii*-caused brown blotch disease in the mushroom context. In addition, the specific mechanism of action of CAR against *P. tolaasii* also has not been elucidated. Several studies indicate that the primary mechanism of CAR in vitro is the disruption of the membrane integrity of bacterial and fungal pathogenic cells, such as *Bacillus cereus* [26], *Listeria monocytogenes* [19], *Aspergillus flavus* [25], and *Colletotrichum fructicola* [27]. Other studies showed that CAR treatment also inhibited respiratory activity in *L. monocytogenes* [19], caused mitochondrial depolarization and DNA damage in *A. flavus* [25], and generated reactive oxygen species (ROS) in *C. fructicola* [27] and *A. flavus* [25]. The present study, therefore, aims to investigate, for the first time, the efficacy of CAR against *P. tolaasii* in vitro and in postharvest *A. bisporus*. In addition, the possible mechanisms of CAR in vitro and in vivo were explored. It is anticipated that the findings of this study will provide a theoretical basis for the further development of postharvest preservation methods for edible mushrooms and the potential application of CAR in the field of edible mushroom preservation.

2. Materials and Methods

2.1. Bacterial Pathogens

The *P. tolaasii* was inoculated into the beef extract–peptone (NA) medium and incubated at 28 °C for 24 h, activated twice, and set aside. *P. tolaasii* colonies, after activation, were selected and cultured in 100 mL of the NB medium (NA without agar), incubated at 28 °C with shaking at 220 rpm until the logarithmic phase was reached.

2.2. In Vitro Antibacterial Activity of CAR Against P. tolaasii

2.2.1. Inhibitory Effect of CAR Treatment on P. tolaasii Growth

P. tolaasii cultures in the logarithmic growth phase were collected and diluted in an NB medium at a ratio of 1:1000. The prepared bacterial suspensions were treated with 1, 2, 3, 4, 5, 6, and 7 mmol L^{-1} of the CAR aqueous solution, respectively, with sterile water as a control (0 mmol L^{-1} CAR). The mixtures were then inoculated into a 96-well plate and incubated for 12–16 h at 28 °C and 220 rpm with shaking, followed by the measurement of absorbance at 600 nm. Each concentration was tested in triplicate.

2.2.2. Killing Effect of CAR Treatment on *P. tolaasii* Growth

P. tolaasii cultures in the logarithmic growth phase were collected and diluted in an NB medium at a ratio of 1:1000. Then, different concentrations of the CAR aqueous solution (0.5, 1, 2, 4, and 8 mmol L^{-1}) were used to treat the bacterial suspensions, with sterile water as the control (0 mmol L^{-1} CAR). Subsequently, the prepared bacterial suspensions were added to a 96-well plate and shaken at 28 °C and 220 rpm for 2 h. The resulting bacterial cultures were then plated on an NA medium and incubated while inverted at 28 °C for 12–16 h. Bacterial colony counts were then determined.

2.2.3. Killing Kinetics of CAR Treatment Against *P. tolaasii*

The killing kinetics of CAR against *P. tolaasii* were evaluated by the method improved by Song et al. [28] The collected logarithmic growth phase of *P. tolaasii* was diluted with the NB medium at a ratio of 1:1000 and subsequently treated with 2, 4, and 8 mmol L^{-1} CAR aqueous solutions, with sterile water serving as the control (0 mmol L^{-1} CAR). After thorough mixing, the suspensions were cultured at 28 °C and 220 rpm for 5, 15, 30, 60, and 120 min. Subsequently, equal volumes of the suspensions were plated on NA agar plates and incubated while inverted at 28 °C for 12–16 h before colony counting.

2.3. *In Vitro Antibacterial Mechanism of CAR Against P. tolaasii*

2.3.1. Effect of CAR Treatment on *P. tolaasii* Cell Morphology

Scanning electron microscopy (SEM) was used to observe changes in cell morphology in CAR-treated *P. tolaasii*, following Zhang et al. [29], with minor adaptations. *P. tolaasii* cells in the logarithmic phase were collected, centrifuged to remove the culture medium, resuspended in PBS solutions containing different concentrations of CAR (0, 2, 4, 8 mmol L^{-1}), and cultured in a shaking incubator at 28 °C and 220 rpm for 2 h. The bacterial cells were then collected by centrifugation and fixed with 4% glutaraldehyde at 4 °C overnight. After fixation, the samples were dehydrated through a series of ethanol gradients [30%, 50%, 70%, 90%, 100% (v/v)] for 5 min each and finally resuspended in 100% ethanol. The ethanol-resuspended bacterial cells were dropped onto the SEM sample stubs, air-dried, coated, and observed for cell morphology.

2.3.2. Effect of CAR Treatment on *P. tolaasii* Cell Membrane Integrity

The propidium iodide (PI) uptake assay was performed according to a previously established method [28] to assess the effect of CAR on *P. tolaasii* membrane integrity. CAR (0, 2, 4, 8 mmol L^{-1}) was used to treat *P. tolaasii* cells, and harvested cells were resuspended in 50 mmol L^{-1} PBS (pH 7.2), mixed with 5 μmol L^{-1} PI and then incubated for 30 min at 28 °C in the dark. Afterward, the cells were observed and imaged with a fluorescence microscope. PBS-treated cells were used as a negative control and were treated and observed under the same conditions.

2.3.3. Effect of CAR Treatment on Cytoplasmic Leakage of *P. tolaasii*

Cytoplasmic leakage was determined according to a previous method [30] using slight modifications. The bacterial suspensions were treated with different concentrations of CAR (0, 2, 4, 8 mmol L^{-1}) at 28 °C and 220 rpm. After treatment for 0, 20, 40, 60, 80, 100, and 120 min, respectively, the supernatants were then collected for nucleic acid, protein, and total sugar measurements. Nucleic acid leakage was assessed by measuring absorbance at 260 nm, protein leakage by the Bradford method [31], and total sugar content by the anthrone–sulfuric acid method [32]. Each sample was used for analysis in triplicate.

2.3.4. Effect of CAR Treatment on the Changes in *P. tolaasii* Cellular Proteins and DNA

As outlined in Section 2.3.1, bacterial suspensions were subjected to varying concentrations of CAR (0, 2, 4, 8 mmol L^{-1}) at 28 °C for 2 h with agitation at 220 rpm, after which they were rinsed with physiological saline solution. SDS-PAGE and Coomassie Brilliant

Blue staining were then conducted, and imaging was performed using the Champ Gel 5000 system (Surwit Technology Co., Ltd., Hangzhou, China) for visualization.

The DNA extraction of *P. tolaasii* cells was conducted using the DNA Extraction Kit (Solarbio, Beijing, China). The extracted DNA was then treated with varying concentrations of CAR (0, 2, 4, 8 mmol L^{-1}) in a 1:1 ratio, followed by a three-hour incubation at 28 °C. The samples were finally subjected to 0.8% (w/v) agarose gel electrophoresis, which was then imaged for observation.

2.4. *Effect of Postharvest CAR Treatment on the Induction of Resistance to Brown Blotch Disease in A. bisporus Mushrooms*

2.4.1. Treatment of *A. bisporus* Mushrooms

Fresh *A. bisporus* mushrooms were procured from Tianshui Zhongxing Bio-Technology Co., Ltd. (Lanzhou, Gansu, China). Following pre-cooling at 4 °C, 600 mushrooms of uniform size, devoid of cap opening, browning, mechanical damage, and disease, were selected. The mushroom stipes were cut evenly to a uniform length and washed with distilled water before air-drying. Subsequently, *A. bisporus* was inoculated with *P. tolaasii*, as previously described [33], and the mushrooms were randomly divided into three groups. According to our preliminary tests, CAR treatments with a concentration of $\geq$30 µmol L^{-1} could cause the white mushroom to turn yellow, so CAR with concentrations of 10 and 20 µmol L^{-1} was selected for mushroom treatments. Specifically, the three groups of mushrooms were covered with a preservation paper wrap that had been soaked with 0 (control), 10, and 20 µmol L^{-1} of CAR solution, respectively, and packaged in a plastic film. The three treatment groups of *A. bisporus* (each with three replicates) were stored under conditions of 85–95% relative humidity and 4 °C temperature. At 0, 12, 24, 36, 48, 60, 72, and 84 h, the number and index of mushrooms developing disease were recorded to calculate the incidence rate and disease index according to the previous method [33], and seven mushrooms were randomly selected and sampled for subsequent analysis in the experiment. The cap lesion tissue, approximately 2 mm thick, was excised and chopped into small pieces to determine the melanin content; the adjacent healthy tissue, approximately 7 mm thick, was immediately exercised and prepared for the determination of the other indicators.

2.4.2. Measurement of MDA, Melanin, Lignin, Total Phenolic, and Flavonoid Content

The MDA content of the mushrooms was determined using the thiobarbituric acid (TBA) method [34]. One gram of the mushroom samples was ground with 5 mL of trichloroacetic acid (10% v/v), centrifuged at 4 °C and 13,000× g for 15 min, and the supernatants were mixed with trichloroacetic acid containing 0.5% (w/v) TBA (10% v/v). Subsequently, the mixture was incubated in a boiling water bath for 20 min, followed by centrifugation to collect the supernatant. Absorbance was measured at wavelengths of 532 nm, 600 nm, and 450 nm to calculate the MDA content.

The melanin content of the mushrooms was analyzed following a method previously described by Selvakumar et al. [35] and Sun et al. [36] with some modifications. One gram of mushroom samples was thoroughly ground with 3 mL of 1 mol L^{-1} NaOH, then subjected to high-pressure sterilization at 121 °C for 20 min. Subsequently, the mixture was centrifuged at 8000× g for 5 min to collect the supernatants, which were subsequently adjusted to a pH of 2.0 by the addition of 7 mol L^{-1} HCl. Following centrifugation at 8000× g for 5 min and the removal of the supernatants, the precipitates were collected and then mixed with 7 mol L^{-1} HCl and placed in a boiling water bath for 20 min. Following centrifugation at 8000× g for 5 min, the melanin precipitates were washed three times with sterile water, air dried, and finally dissolved in 1 mol L^{-1} NaOH for the absorbance measurement at 400 nm. Each sample was analyzed in triplicate. A standard curve for melanin was concurrently plotted to calculate the melanin content in the samples based on the equation of the standard curve.

The lignin content of the mushrooms was determined by means of a previously established method [37]. One gram of the sample was ground in 2.0 mL of 95% ethanol solution and centrifuged at 8000× *g* for 7 min at 4 °C. Subsequently, the supernatants were discarded, and the resulting precipitate was washed three times with 95% ethanol and ethanol–hexane (1:2, *v*/*v*), respectively. It was then redissolved and incubated in a water bath at 70 °C for 30 min in 0.5 mL of 25% (*v*/*v*) bromoacetyl solution. We then added 0.9 mL of 2 mol L^{-1} NaOH to terminate the reaction. Subsequently, 5 mL of glacial acetic acid and 0.1 mL of 7.5 mol L^{-1} hydroxylamine hydrochloride were added, and the reaction was centrifuged at 8000× *g* for 7 min. Finally, the absorbance of the supernatant was measured at 280 nm. The lignin content in each sample was calculated from the standard curve of the lignin standard and expressed as grams per kilogram of fresh weight (g kg^{-1} FW). The assay was repeated three times for each sample.

The total phenolic and flavonoid content of the mushrooms was determined in accordance with the methodology described in Yang et al. [33], with certain modifications. The sample tissues were ground with a 1% (*v*/*v*) hydrochloric acid–methanol solution in an ice bath and incubated at 4 °C for 2 h in the dark. Subsequently, the supernatants were centrifuged at 8000× *g* for 15 min at 4 °C. Subsequently, absorbance measurements at 280 nm and 325 nm were carried out, and standard curves for gallic acid and catechin were used to calculate the total phenolic and flavonoid contents (g kg^{-1} FW), respectively.

2.4.3. Enzyme Activity Assays

Peroxidase (POD), superoxide dismutase (SOD), polyphenol oxidase (PPO), and catalase (CAT) were measured using assay kits A084-3-1, A001-1-1, A136-1-1, and A007-1-1, respectively, which were purchased from Nanjing Jiancheng Bioengineering Institute (Nanjing, Jiangsu, China).

To extract β-1,4-N-acetyl-glucosaminnidase (NAG), 1.0 g of the mushroom samples was thoroughly ground in 3 mL of a 100 mmol L^{-1} acetic acid–sodium acetate buffer (pH 5.0) at 4 °C and centrifuged at 8000× *g* for 15 min. Subsequently, the supernatants were mixed with a 100 mmol L^{-1} acetic acid-sodium acetate buffer (pH 5.0, containing 300 mg L^{-1} of p-nitrophenyl-N-acetyl-β-D-glucosaminide). After incubation for 30 min at 37 °C, 0.4 mol L^{-1} Na_2CO_3 was added to terminate the reaction, and the absorbance of the mixture was measured at 405 nm. Each sample was measured three times. The amount of p-nitrophenol generated was measured from the standard curve, and NAG activity was defined as micromoles per hour per milligram of protein (μmol h^{-1} mg^{-1} protein).

The extraction of phenylalanine ammonia-lyase (PAL) was conducted using a 200 mmol L^{-1} borate buffer (pH 8.8), and the activity was determined according to the method of Ji et al. [38] One unit (U) of PAL activity represents a change in absorbance at 290 nm of 1.0 per hour.

The extraction of cinnamic 4-coumarate-CoA ligase (4CL) and acid-4-hydroxylase (C4H) was conducted using a 50 mmol L^{-1} Tris-HCl buffer (pH 8.9). Subsequently, the 4CL and C4H activity assays were performed in accordance with the methodologies described by Ji et al. [38] and Wang et al. [39], respectively. One unit (U) of C4H activity represents a 0.001 change in absorbance at 340 nm per hour, while one unit (U) of 4CL activity represents a 0.01 change in absorbance at 333 nm per hour.

The extraction of cinnamyl alcohol dehydrogenase (CAD) was conducted using a 0.2 mmol L^{-1} Tris-HCl buffer (pH 6.25), and the activity was determined according to the method of Li et al. [40]. One unit (U) of CAD activity represents 0.01 of change in absorbance at 340 nm per hour.

2.5. Statistical Analysis

The experimental data were subjected to statistical analysis using analysis of variance (ANOVA) with IBM SPSS Statistics version 25.0.

3. Results

3.1. CAR Treatment Showed a Strong Inhibitory and Killing Effect on the Growth of P. tolaasii In Vitro

As shown in Figure 1A, CAR treatment significantly inhibited the growth of *P. tolaasii*, and increasing concentrations of CAR resulted in an increasing inhibitory effect. When the concentration of CAR reached 4 and 5 mmol L^{-1}, the inhibition rate exceeded 90% and 99%, respectively. Figure 1B shows the killing effect of different concentrations of CAR on *P. tolaasii*. When the concentration of CAR increased to 8 mmol L^{-1}, no bacterial colonies were observed on the plate, indicating that 100% of *P. tolaasii* was effectively killed. To further verify the killing efficiency of CAR on *P. tolaasii*, the killing kinetic assays were performed (Figure 1C). The control group showed a continuous increase in colony count over the incubation period, while the CAR-treated groups showed a decreasing trend. In particular, 4 and 8 mmol L^{-1} CAR showed a rapid killing effect during the first 5 min, and 8 mmol L^{-1} CAR was able to kill all *P. tolaasii* within 5 min. These results indicate that CAR has a potent inhibitory and killing effect against *P. tolaasii* in vitro.

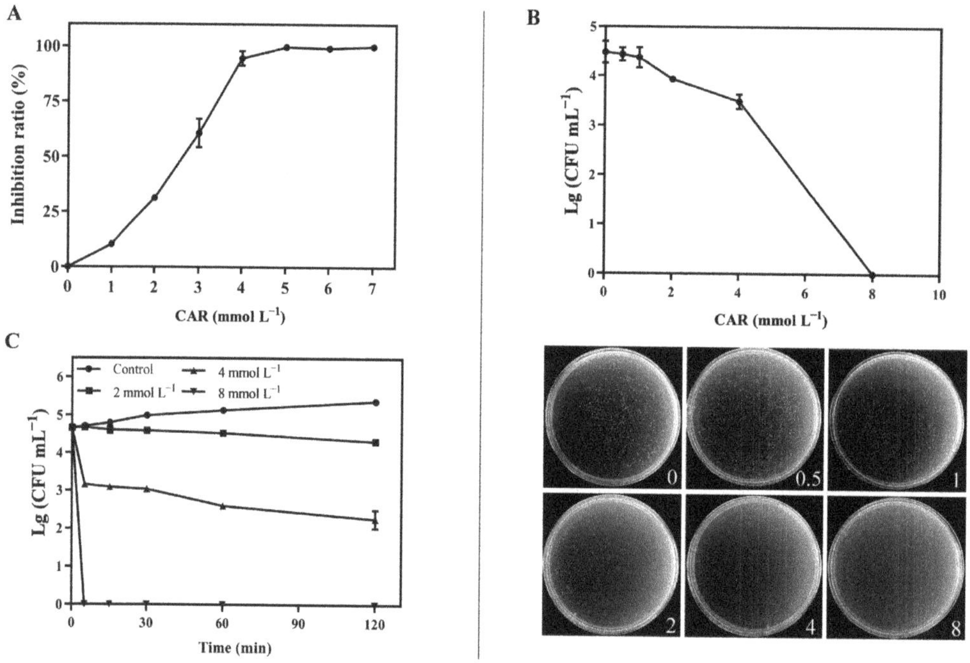

Figure 1. Antibacterial activity of CAR against *P. tolaasii* in vitro. (**A**) The inhibition ratio of different concentrations of CAR on *P. tolaasii* growth. (**B**) Killing effect of different concentrations of CAR against *P. tolaasii*. (**C**) Killing kinetics of CAR against *P. tolaasii*. Data represent the means ± SD, n = 3.

3.2. CAR Treatment Altered the Cellular Morphology of P. tolaasii

The SEM observation of the cellular morphology of *P. tolaasii* revealed that different concentrations of CAR caused different degrees of damage to the cellular morphology of *P. tolaasii* compared to the control group (Figure 2). Untreated *P. tolaasii* cells appeared plump with a smooth surface. Treatment with 2 mmol L^{-1} CAR resulted in some *P. tolaasii* cells losing their basal shape and exhibiting fragmentation, adhesion, and aggregation. As the concentration of CAR increased, the degree of cell fragmentation also increased. When the treatment concentration reached 8 mmol L^{-1}, *P. tolaasii* completely lost its cellular morphology, and the cells were completely fragmented and adherent. These results indicate the destructive effect of CAR on *P. tolaasii* cells.

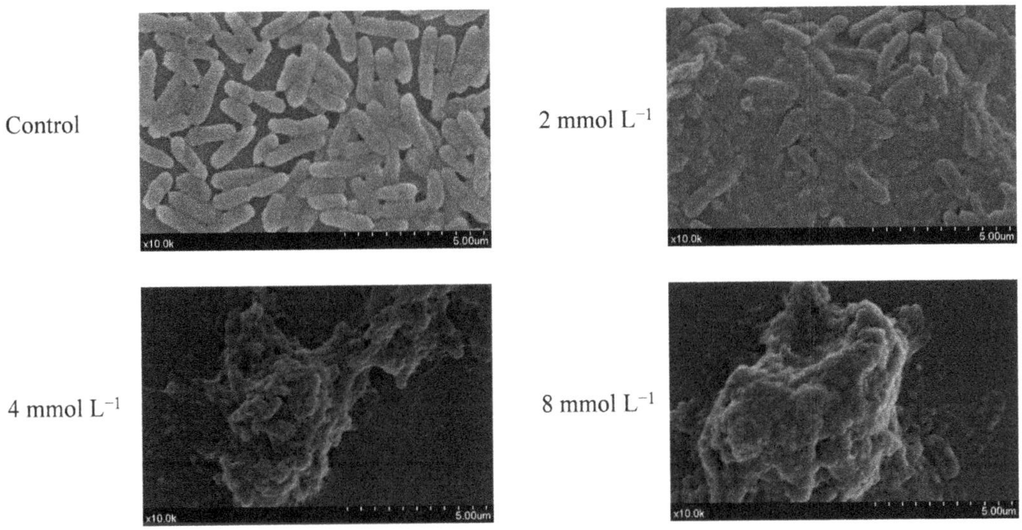

Figure 2. Effect of CAR treatment on the cell morphology of *P. tolaasii* in the 0 (control), 2, 4 or 8 mmol L^{-1} of CAR treatment groups.

3.3. CAR Treatment Disrupts the Membrane Integrity of *P. tolaasii*

Based on the SEM observation results, *P. tolaasii* was further subjected to PI staining and fluorescence signal detection. PI cannot penetrate intact cell membranes for staining, but it can pass through damaged membranes, bind to the DNA inside the cell, and show red fluorescence. As shown in Figure 3A, *P. tolaasii* treated with PBS showed no red fluorescence signal, while the three groups treated with different concentrations of CAR showed an obvious red fluorescence signal. In particular, 8 mmol L^{-1} CAR showed the strongest signal compared to the bacterial cells treated with 2 mmol L^{-1} and 4 mmol L^{-1} CAR. These results suggest that CAR treatment can disrupt the integrity of *P. tolaasii* bacterial cell membranes, and the degree of disruption increases with higher concentrations of CAR treatment. We also evaluated cell membrane damage by measuring the leakage of nucleic acids, proteins, and soluble carbohydrates within the *P. tolaasii* cells (Figure 3B). Higher absorbance values at 260 nm and higher levels of soluble protein and total sugars were consistently detected in the supernatants of cells treated with 2, 4, and 8 mmol L^{-1} CAR compared to cells without CAR treatment (control) at all time points over a 120 min period. In addition, OD260, a soluble protein, and total sugar levels were increased by CAR treatment in a concentration-dependent manner, with 8 mmol L^{-1} CAR resulting in the highest increase. Taken together, our results demonstrate that CAR treatments can disrupt the integrity of *P. tolaasii* cell membranes, resulting in the leakage of cytoplasmic constituents.

3.4. CAR Treatment Did Not Affect the Intracellular Proteins and DNA of *P. tolaasii*

To investigate whether CAR acts against pathogens by interacting with the intracellular proteins or DNA of *P. tolaasii*, we next performed SDS-PAGE analysis and DNA binding assays. As shown in Figure 4A, there were no clear changes in the protein bands after different concentrations of CAR treatments, indicating that CAR treatments did not affect the synthesis and stability of intracellular proteins. Similarly, the gel retardation assays showed that all the CAR treatments also had no significant effect on the *P. tolaasii* DNA migration (Figure 4B), indicating that CAR could not interact with or degrade *P. tolaasii* DNA.

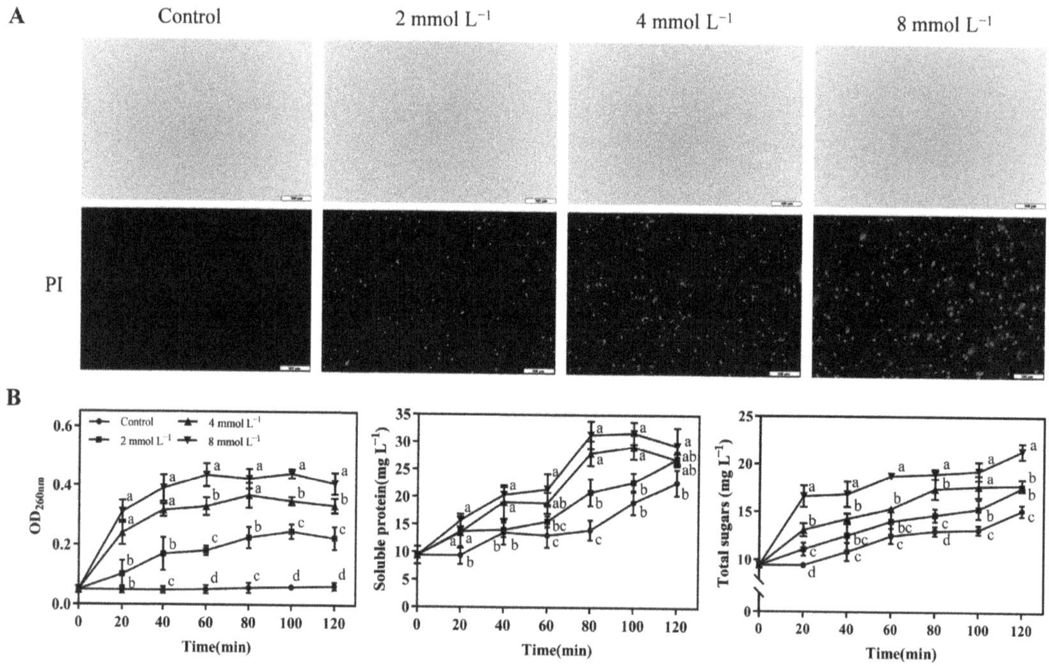

Figure 3. Effect of CAR treatment on the plasma membrane of *P. tolaasii* cells. (**A**) Fluorescent microscope observations of PI uptake into *P. tolaasii* cells in 0 (control), 2, 4 or 8 mmol L^{-1} of the CAR treatment groups. (**B**) The leakage of cytoplasmic leakage of nucleic acid (absorbance at 260 nm), soluble proteins, and total sugars from *P. tolaasii* in 0 (control) at 2, 4, or 8 mmol L^{-1} of CAR treatment groups during the 120 min duration. Data represent the means ± SD, $n = 3$. Different lowercase letters indicate groups with significant differences at $p < 0.05$.

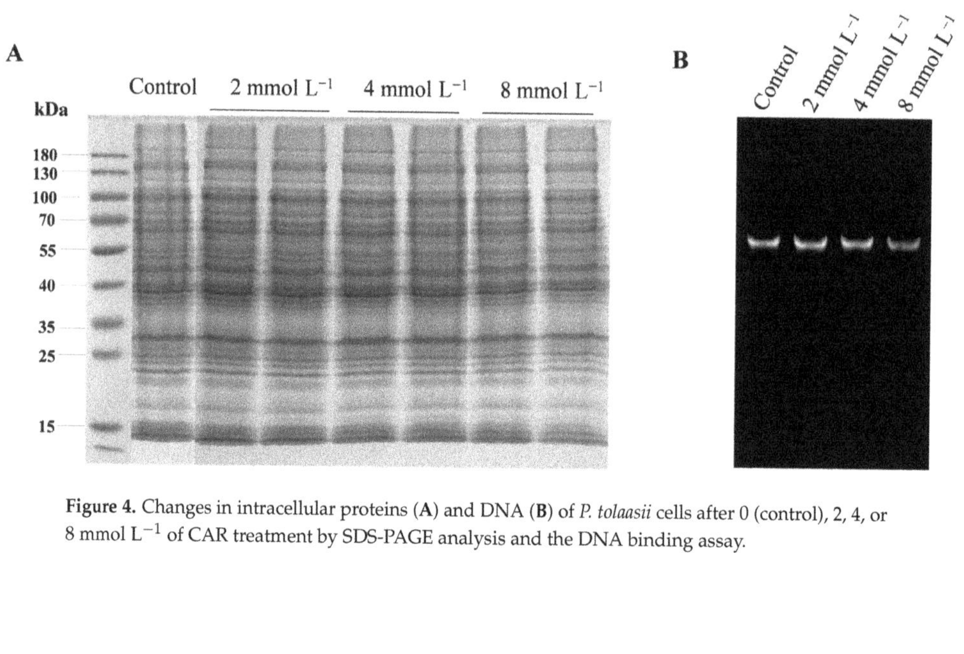

Figure 4. Changes in intracellular proteins (**A**) and DNA (**B**) of *P. tolaasii* cells after 0 (control), 2, 4, or 8 mmol L^{-1} of CAR treatment by SDS-PAGE analysis and the DNA binding assay.

3.5. CAR Treatment Inhibited *P. tolaasii*-Caused Brown Blotch Disease in *A. bisporus* Mushrooms

To determine the effect of CAR treatments on the development of bacterial brown blotch disease in edible mushrooms, the sporophores of *A. bisporus* were treated with CAR. The control group showed disease spots 36 h after inoculation, and the spots became progressively larger and darker as the storage time increased (Figure 5). In contrast, the mushrooms treated with 10 and 20 μmol L^{-1} CAR began to show clear symptoms of the disease at 48 h post-inoculation, with the severity of the disease being significantly less in the 20 μmol L^{-1} CAR treatment group than in the 10 μmol L^{-1} treatment group (Figure 5A). Furthermore, both concentrations were found to reduce the incidence and index of disease during storage. Notably, the 20 μmol L^{-1} CAR-treated group showed the lowest disease incidence and index and achieved the most effective disease control (Figure 5B). The incidence rate of the control group mushrooms reached a 100% incidence rate at 60 h, while the 10 μmol L^{-1} and 20 μmol L^{-1} treatment groups were only 91.4 % and 82.5%, respectively. After 60 h of storage, the disease index in the control group was approximately 1.55 times and 1.88 times higher than that in the 10 μmol L^{-1} and 20 μmol L^{-1} groups, respectively. After 72 h of storage, the disease index in the control group exceeded 50%, while the 20 μmol L^{-1} CAR-treated group showed only 37.4%. Collectively, CAR treatment prevented the development of brown blotch disease in *A. bisporus* mushrooms caused by *P. tolaasii*.

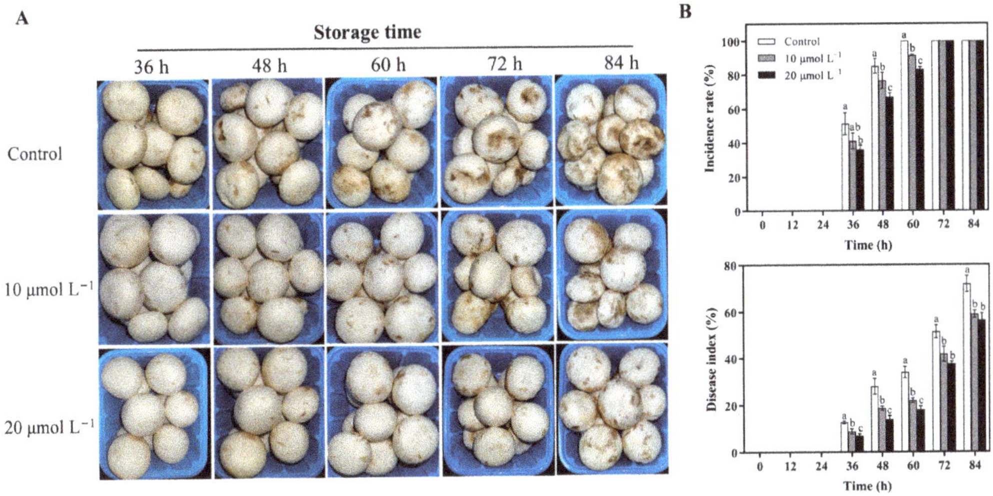

Figure 5. Postharvest treatment with CAR suppressed brown spot disease in mushrooms caused by *P. tolaasii*. The figure shows the visual symptoms (**A**), incidence rate, and disease index (**B**) of mushrooms treated with 0 (control), 10, and 20 μmol L^{-1} CAR during storage. Data represent the means ± SD, $n = 3$. Different lowercase letters indicate groups with significant differences at $p < 0.05$.

3.6. CAR Treatment Inhibited *P. tolaasii*-Caused PPO Activation and Melanin Production and Reduced MDA Accumulation in *A. bisporus* Mushrooms

The PPO catalyzes the oxidation of phenolic compounds in living organisms to quinones, resulting in browning [41,42]. Tolaasin secreted by *P. tolaasii* can disrupt the integrity of cell membranes and activate PPO, resulting in the appearance of brown lesions on the surface of infected mushrooms [43,44]. Throughout the storage period, the trend of PPO activity showed an initial increase followed by a decrease, as shown in Figure 6A. In the control group, PPO activity increased sharply after 12 h and peaked at 24 h. However, both the 10 μmol L^{-1} and 20 μmol L^{-1} CAR treatment groups delayed the peak PPO activity to 36 h. It is noteworthy that the peak value in the 10 μmol L^{-1} and 20 μmol L^{-1} CAR treatment groups was only 80% and 72% of that in the control group, respectively.

In addition, the PPO activity in the 20 μmol L^{-1} CAR treatment group was the lowest at almost all the time points.

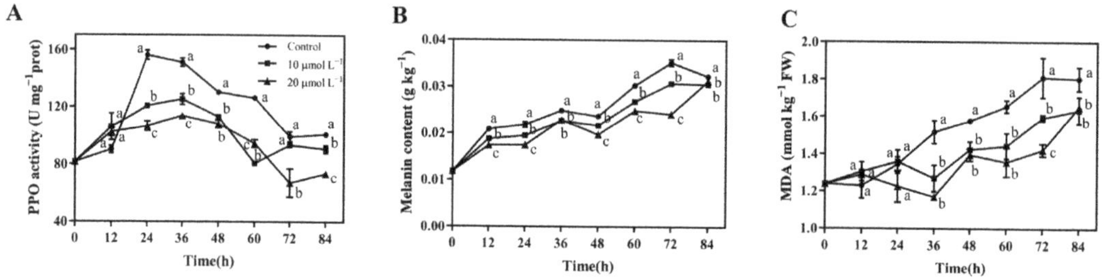

Figure 6. Effects of postharvest CAR treatments on PPO activity (**A**), melanin content (**B**), and MDA content (**C**) in *A. bisporus* after inoculation with *P. tolaasii*. Data represent the means ± SD, $n = 3$. Different lowercase letters indicate groups with significant differences at $p < 0.05$.

Melanin content is a product of the browning reaction that can visually reflect the degree of disease development in *A. bisporus*. Figure 6B shows that with a prolonged storage time, the melanin content gradually increased in all three treatment groups. Consistent with PPO activity, the melanin content in the 10 μmol L^{-1} and 20 μmol L^{-1} CAR treatment groups was also consistently lower than the control group throughout the storage, with the lowest values in the 20 μmol L^{-1} CAR treatment group. The above results indicate that CAR treatment could inhibit the *P. tolaasii*-induced PPO activation and, hence, melanin production in *A. bisporus* mushrooms.

In addition to melanin, the MDA content also showed an overall increasing trend during the storage process, reaching its peak at 84 h, indicating the continuous accumulation of MDA in *A. bisporus* mushrooms. In the control group, the MDA content exhibited a marked increase from 12 h onwards, reaching levels 1.28 and 1.47 times higher than the initial value after 48 and 72 h of storage, respectively. In contrast, only a very slight increase in MDA content was detected in the CAR treatment groups during the first 48 h of storage. Therefore, CAR treatment could also inhibit the MDA accumulation in *A. bisporus* mushrooms.

3.7. CAR Treatment Increased CAT, POD, SOD, and NAG Activity in Postharvest A. bisporus to Resist P. tolaasii Infection

Our previous research has shown that the enhancement of antioxidant enzyme activities is beneficial for the postharvest resistance of mushrooms to bacterial brown blotch disease [33]. Hence, to investigate whether CAR could induce the disease resistance of *A. bisporus* mushrooms against bacterial brown blotch disease, we performed the measurements and analysis of antioxidant enzyme activities, including CAT, POD, and SOD. Figure 7A–C show that within 84 h of storage, the CAR-treated groups had consistently higher activities of CAT, POD, and SOD than the control group, with the 20 μmol L^{-1} CAR group being higher than the 10 μmol L^{-1} CAR group. From 12 to 48 h after inoculation, the CAT activity in the 20 μmol L^{-1} CAR treatment group was found to be approximately 1.33 to 2.24 times higher than that in the control group. Similarly, the POD and SOD activities in the 20 μmol L^{-1} CAR treatment group were 1.46 to 2.23 and 1.18 to 1.42 times higher, respectively, than those in the control group at 12 to 60 h post-inoculation.

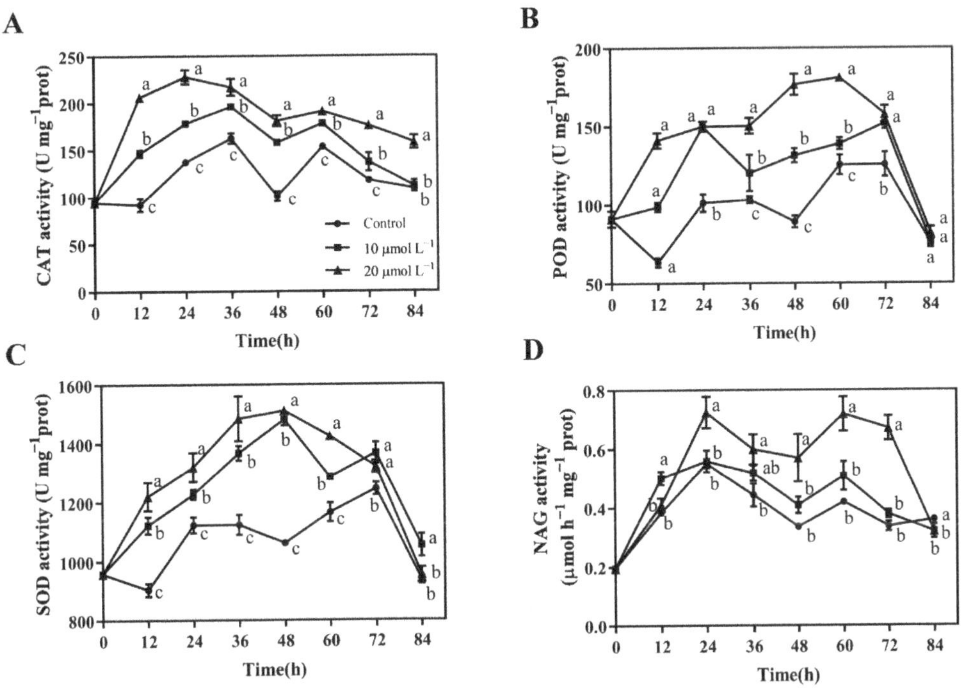

Figure 7. Effects of postharvest CAR treatments on the activities of CAT (**A**), POD (**B**), SOD (**C**), and NAG activity (**D**) in *A. bisporus* after inoculation with *P. tolaasii*. Data represent the means ± SD, $n = 3$. Different lowercase letters indicate groups with significant differences at $p < 0.05$.

Since *A. bisporus* could secrete NAG, which is an extracellular enzyme capable of degrading bacteria, we next investigated whether CAR treatment could induce NAG to degrade *P. tolaasii*. Figure 7D shows that NAG activity remained higher in the CAR-treated groups than in the control group from 24 h to 72 h during storage. At 60 h, the NAG enzyme activity in the 10 and 20 μmol L^{-1} CAR-treated groups was 1.21 and 1.71 times higher than that in the control group, respectively. The aforementioned results indicated that CAR treatment was effective in inducing both antioxidant enzymes and the bacterial degrading enzyme for defense against *P. tolaasii* infection.

3.8. CAR Treatment Promoted the Phenylpropanoid Pathway in Postharvest A. bisporus to Resist P. tolaasii Infection

Secondary metabolites (e.g., phenolics and flavonoids) produced in the phenylpropanoid pathway play a pivotal role in plant responses to both biotic and abiotic stressors [45]. Therefore, the effect of CAR treatment on the phenylpropanoid pathway in postharvest *A. bisporus* mushrooms was investigated in the current study. As illustrated in Figure 8, the activities of key enzymes (including PAL, C4H, 4CL, and CAD) in the phenylpropanoid pathway were all markedly elevated by CAR treatments during the 84 h storage period in comparison to the control group, resulting in a notable increase in the levels of total phenolics, flavonoids, and lignin. Furthermore, with respect to all indicators depicted in Figure 8, CAR at 20 μmol L^{-1} exhibited a more pronounced promoting effect than the 10 μmol L^{-1} CAR treatment on the phenylpropanoid pathway.

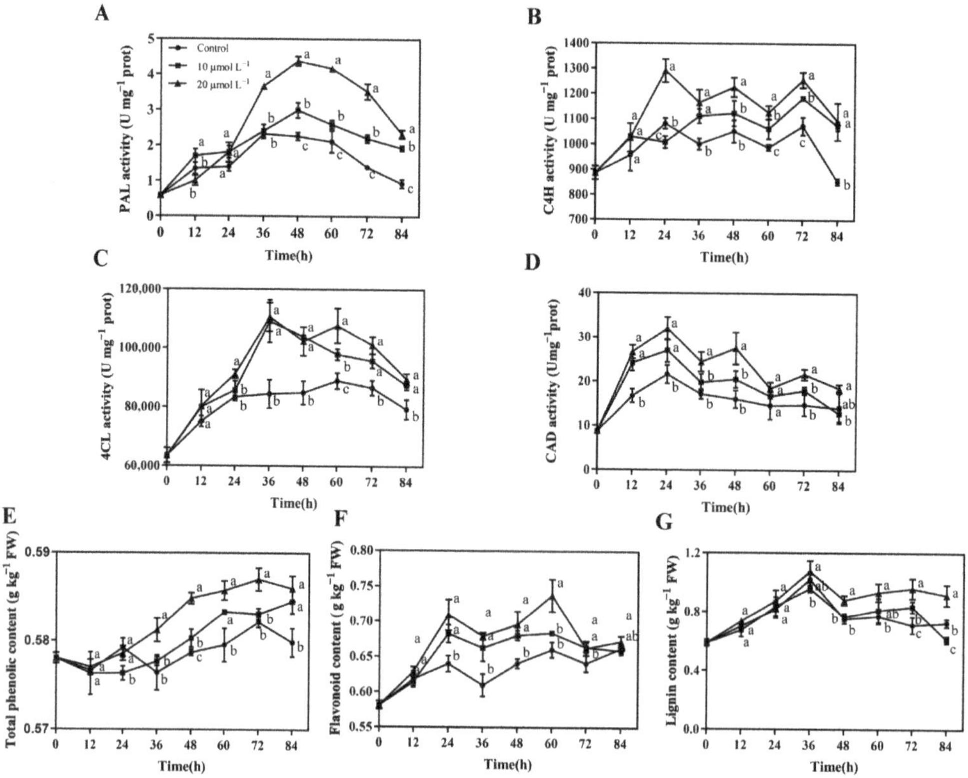

Figure 8. Effects of postharvest CAR treatments on the activities of PAL (**A**), C4H (**B**), 4CL (**C**), and CAD (**D**), and the contents of total phenolics (**E**), flavonoids (**F**), and lignin (**G**) in *A. bisporus* after inoculation with *P. tolaasii*. Data represent the means ± SD, $n = 3$. Different lowercase letters indicate groups with significant differences at $p < 0.05$.

Specifically, CAR treatments significantly promoted the increase in PAL activity during the second half of the storage period (36–84 h), with the peak value in the 20 μmol L^{-1} CAR-treated groups being 1.95 times higher than that of the control group (Figure 8A). CAR treatments led to a significant increase in C4H activity from 24 h onwards, with the activity in the 20 μmol L^{-1} CAR-treated group being 113% to 128% of that in the control group (Figure 8B); 4CL activity in the 10 μmol L^{-1} and 20 μmol L^{-1} CAR treatments were approximately 1.03–1.31 times higher than that of the control from 24 h to 84 h (Figure 8C). CAD activity was rapidly elevated by CAR treatments during the first half of storage (12–48 h), and the activity in the 20 μmol L^{-1} CAR-treated group at 12 and 24 h was approximately 1.60 and 1.48 times higher than that of the control, respectively (Figure 8D).

It was observed that total phenolics and flavonoids were produced and accumulated during the defense process of *A. bisporus* against *P. tolaasii* (Figure 8E,F). Following inoculation, the total phenolic content in the 20 μmol L^{-1} CAR-treated group was approximately 1.01 times higher than that of the control group over the 36–84 h that followed. Similarly, the total flavonoid content in the 20 μmol L^{-1} CAR-treated group was approximately 1.09–1.12 times higher than that of the control group during the 24 h to 60 h post-inoculation period (Figure 8E,F). In comparison to the control group, the lignin content of the 10 μmol L^{-1} CAR-treated group exhibited a significant increase only at 72 h of storage (Figure 8G). In contrast, the 20 μmol L^{-1} CAR-treated group demonstrated a significantly higher lignin content than the control group from 48 h onwards, with the greatest increase observed being 135% of the control at 72 h. It can be concluded that the CAR treatment activated the phenylpropanoid pathway, which may contribute to the resistance of *A. bisporus* to infection by *P. tolaasii*. This was evidenced by the elevated activities of associated enzymes and the accumulation of metabolic substances.

4. Discussion

Brown blotch disease, caused by the bacterial pathogen *P. tolaasii*, is a key factor leading to the postharvest deterioration of edible mushrooms, and currently, brown blotch disease has been identified in various edible mushrooms [46]. CAR, as a promising environmentally friendly antimicrobial substance, has recently received widespread attention for its potential use in food preservation [47]. However, to date, the antimicrobial activity of CAR against the pathogen *P. tolaasii* in vitro and in edible mushrooms has not been investigated.

The present study first demonstrated that CAR not only has a strong inhibitory effect on the growth of *P. tolaasii* but also can completely kill the pathogens within 5 min when 8 mmol L^{-1} of CAR is used (Figure 1). This result is consistent with previous reports, which documented that the minimum inhibitory concentration (MIC) and minimum bactericidal/fungicidal concentration (MBC/MFC) of CAR against bacterial and fungal pathogens range from 78 to 1250 mg L^{-1} [12,48,49]. Notably, subsequent observations revealed that the concentration of CAR required to inhibit the brown blotch disease of *A. bisporus* caused by *P. tolaasii* was considerably lower than that required in the in vitro experiments. A 20 μmol L^{-1} CAR treatment was observed to significantly inhibit the development and expansion of the brown spots on the cap of the mushrooms, as well as reduce the incidence rate and disease index (Figure 5). Treatments with a concentration of $\geq$ 30 μmol L^{-1} were observed to result in the white mushroom turning yellow and exacerbating the disease. Previous researchers have also highlighted the potential for CAR with a high concentration to have adverse effects on fruits and phytotoxic effects on crop plants due to its characteristics of small particle size and effective penetration capability [23,50]. Therefore, the utilization of an appropriate, yet not excessive, concentration of CAR is advised for the control of mushroom diseases in practical applications.

Given the inherently volatile nature of CAR and the necessity to minimize the adverse effects of CAR on mushroom tissues, a fumigation method was employed in the current study for the treatment of mushrooms with CAR. The inhibitory effects of CAR on *Aspergillus flavus* infection during wheat storage were also previously investigated by Duan et al. [25] using a fumigation method. Following CAR fumigation, a significant reduction in disease incidence and index was observed, particularly in the period preceding 60 h post-inoculation. In the 20 μmol L^{-1} CAR-treated mushrooms, the disease index remained consistently below 55% of the control value (Figure 5). At 72 and 84 h, the disease index reached a value of 72.8 and 78.6% in the control group, respectively, indicating a gradual decline in the efficacy of CAR in inhibiting mushroom disease as time progressed. To enhance the stability and residence time of CAR, several researchers have recently explored the incorporation of CAR onto specific carriers [13,51,52], the encapsulation of CAR using stabilizers [53], and the development of active packaging utilizing CAR as a component [54]. For example, Sánchez-Hernández et al. [13] discovered that the chitosan-carboxymethylcellulose-alginate nanocarriers encapsulated with CAR required lower concentrations of CAR to achieve the same inhibitory effect. The MIC values for the encapsulated CAR treatments were 23.3 to 31.3 μg mL^{-1} against *Botrytis cinerea*, *Penicillium expansum*, and *Colletotrichum coccodes*, whereas the unencapsulated CAR treatments exhibited MIC values of 500 to 1500 μg mL^{-1}. Consequently, a practical dose of CAR (50–100 μg mL^{-1}) was found to provide complete protection in postharvest tomatoes against pathogens, effectively resolving the controversy surrounding the application of high doses of CAR in crop protection. Similarly, Fang et al. [54] reported that the combination of CAR with a whey protein isolate emulsion and ε-polylysine (ε-PL) had a beneficial effect on the storage stability of CAR. The prepared CAR/ε-PL-loaded nanoemulsion was effective in suppressing the formation of *Staphylococcus aureus* and *Escherichia coli* biofilms and in preserving mangoes from black rot during a seven-day storage period. Therefore, further research is required to explore different forms or combinations of CAR utilization, such as encapsulating and combination methods, to achieve a more effective control effect in the preservation of edible mushrooms.

It has been shown previously that CAR can exert antibacterial effects by altering the permeability of bacterial membranes [12]. Therefore, in the current study, the effects of CAR on the cytoplasmic membrane of *P. tolaasii* were also investigated. As shown in Figure 3, CAR treatments induced the disruption of *P. tolaasii* membranes accompanied by the leakage of vital intercellular components, including nucleic acids, soluble proteins, and soluble carbohydrates, consequently resulting in significant damage to the cellular morphology of *P. tolaasii* (Figure 2). Furthermore, the degree of cell disruption and leakage was in a concentration-dependent manner after CAR treatment. Previous reports also indicated that CAR exhibited similar bactericidal mechanisms against *Pseudomonas aeruginosa* [48] and *Dickeya zeae* [55]. Therefore, it could be inferred that CAR inhibits the growth of *P. tolaasii* by disrupting the plasma membranes and inducing the release of intracellular components. To further investigate whether CAR can target cellular components, we examined the effects of CAR on intracellular proteins and DNA in *P. tolaasii* cells. It was found that CAR treatment did not alter the bands of intracellular proteins and DNA (Figure 4). Yin et al. [56] found that CAR disrupted *Lasiodiplodia theobromae*'s plasma membrane and bound to its DNA, thereby controlling tea leaf disease. This differs from our findings, likely due to the differences in the characteristics of the pathogen itself. In addition, CAR also has other antimicrobial mechanisms, such as inducing endoplasmic reticulum stress [57] and producing reactive oxygen species [23]. Therefore, further in-depth investigations are needed in the future to determine whether CAR can kill *P. tolaasii* by other mechanisms.

It is of particular significance that the present study has demonstrated that the application of CAR has the potential to enhance disease resistance against brown blotch disease in postharvest *A. bisporus* mushrooms. This is achieved by increasing the activities of three antioxidant-related enzymes (CAT, POD, and SOD) and one bacterial degradation-related enzyme (NAG), as well as promoting pathway-related enzyme activities (PAL, C4H, 4CL, and CAD) and the accumulation of phenolic, flavonoid, and lignin substances (Figures 7 and 8). Consequently, CAR treatments effectively inhibited *P. tolaasii*-induced PPO activation and melanin production, as well as MDA accumulation in *A. bisporus* mushrooms (Figure 6). The phenylpropanoid pathway exerts important roles in the defense system and postharvest quality improvement of mushrooms [58–60]. The secondary metabolites of the phenylpropanoid pathway, total phenolics, and flavonoids, are important active substances that exert antibacterial and antioxidant effects [61]. In our previous study, the activities of CAT, POD, and SOD, as well as the phenylpropanoid pathway, were also shown to play an important role in the defense response of *A. bisporus* against *P. tolaasii* [33] and ε-PL-induced resistance [28]. In agreement with our findings, a similar phenomenon was reported in postharvest kiwifruit, where CAR treatment significantly increased CAT, POD, SOD, PAL, C4H, and 4CL activities and increased total phenolic and flavonoid contents, thereby inducing fruit resistance to stem-end rot [62]. Similarly, Li et al. [23] reported that CAR treatment enhanced the resistance of red grapes to the decay disease caused by *Alternaria alternata* during storage by increasing the activities of antioxidant response-related enzymes (CAT, POD, and SOD) and a PAL enzyme. It is well known that chitinase and glucanase are hallmark enzymes that are involved in plant resistance to fungal pathogens due to their ability to directly hydrolyze the fungal cell wall [63,64]. However, there is a paucity of information regarding the pathogenesis-related enzymes associated with resistance to bacterial pathogens. It is worth mentioning that NAG was observed to be significantly induced by CAR treatment against *P. tolaasii* pathogens, indicating, for the first time, that it is an important bacterial defense-related enzyme in edible mushrooms, with the ability to directly degrade bacterial pathogens. Nevertheless, further study is required to determine whether there are other pathogenesis-related proteins involved.

5. Conclusions

The present study demonstrates the potent antibacterial activity of CAR against *P. tolaasii* in vitro and its efficacy in controlling the brown blotch disease of postharvest *A. bisporus* mushrooms caused by *P. tolaasii*. The primary mechanism of action of CAR against *P. tolaasii* is the disruption of cell membranes. Moreover, the current results indicate that exogenous CAR treatment can induce resistance to bacterial brown blotch disease by increasing the activities of antioxidant-related enzymes, a bacterial degradation-related enzyme, and phenylpropanoid pathway-related enzymes, as well as by promoting the accumulation of phenolic, flavonoid, and lignin substances in the mushrooms. Therefore, CAR has promising potential for the control of bacterial brown blotch disease in edible mushrooms. In subsequent studies, we aim to investigate strategies for optimizing the efficacy of CAR in the control of mushroom diseases.

Author Contributions: L.Z.: Validation, Formal analysis, and Writing—original draft. R.S.: Methodology, Data curation and Writing—original draft. Z.S.: Formal analysis and Investigation. S.Y.: Investigation and Validation. L.J.: Formal analysis. M.M.: Validation. X.W.: Validation. L.C.: Visualization. X.L.: Resources. D.M.: Conceptualization, Resources, Supervision, Funding acquisition, and Writing—Review and Editing. All authors have read and agreed to the published version of the manuscript.

Funding: This research was supported by National Natural Science Foundation of China (No. 32372391).

Institutional Review Board Statement: Not applicable.

Informed Consent Statement: Not applicable.

Data Availability Statement: The original contributions presented in the study are included in the article; further inquiries can be directed to the corresponding author.

Conflicts of Interest: The authors declare no conflicts of interest.

References

1. Duan, H.; Yu, Q.; Ni, Y.; Li, J.W.; Yu, L.L.; Fan, L.P. Calcium combined with vacuum treatment improves postharvest storage quality of *Agaricus bisporus* by regulating polyamine metabolism. *Postharvest Biol. Technol.* **2024**, *210*, 112375. [CrossRef]
2. Ramos, M.; Burgos, N.; Barnard, A.; Evans, G.; Preece, J.; Graz, M.; Ruthes, A.C.; Jiménez-Quero, A.; Martínez-Abad, A.; Vilaplana, F.; et al. *Agaricus bisporus* and its by-products as a source of valuable extracts and bioactive compounds. *Food Chem.* **2019**, *292*, 176–187. [CrossRef] [PubMed]
3. Bai, S.; Han, P.; Zhou, Z.; Ma, N.; Fang, D.; Yang, W.; Hu, Q.; Pei, F. Caffeic acid-grafted chitosan/polylactic acid packaging affects bacterial infestation and volatile flavor of postharvest *Agaricus bisporus*. *J. Food Compos. Anal.* **2023**, *122*, 105504. [CrossRef]
4. Faraj, A.M.; Nouri, M. Development of a mucilage coating including nanoencapsulated essential oils for extending shelf life of button mushrooms (*Agaricus bisporus*). *Food Packag. Shelf* **2024**, *41*, 101232. [CrossRef]
5. Jiang, W.; Zhu, D.; Zhao, L.; Liu, Y.; Wang, C.; Farid, M.S.; Gu, Y.; Li, J.; Li, T.; Sun, Y. L-cysteine treatment delayed the quality deterioration of fresh-cut button mushrooms by regulating oxygen metabolism, inhibiting water loss, and stimulating endogenous H_2S production. *J. Agric. Food Chem.* **2022**, *71*, 974–984. [CrossRef] [PubMed]
6. Ghasemi, S.; Harighi, B.; Azizi, A.; Mojarrab, M. Reduction of brown blotch disease and tyrosinase activity in *Agaricus bisporus* infected by *Pseudomonas tolaasii* upon treatment with endofungal bacteria. *Physiol. Mol. Plant Pathol.* **2020**, *110*, 101474. [CrossRef]
7. Castellanos-Reyes, K.; Villalobos-Carvajal, R.; Beldarrain-Iznaga, T. Fresh mushroom preservation techniques. *Foods* **2021**, *10*, 2126. [CrossRef]
8. Kesraoui, S.; Andrés, M.F.; Berrocal-Lobo, M.; Soudani, S.; Gonzalez-Coloma, A. Direct and indirect effects of essential oils for sustainable crop protection. *Plants* **2022**, *11*, 2144. [CrossRef]
9. Yasmeen, N.; Chaudhary, A.A.; Khan, S.; Ayyar, P.V.; Lakhawat, S.S.; Sharma, P.K.; Kumar, V. Antiangiogenic potential of phytochemicals from *Clerodendrum inerme* (L.) Gaertn investigated through in silico and quantum computational methods. *Mol. Divers.* **2024**, 1–25. [CrossRef]
10. de Souza, E.L.; de Barros, J.C.; de Oliveira, C.E.V.; da Conceição, M.L. Influence of *Origanum vulgare* L. essential oil on enterotoxin production, membrane permeability and surface characteristics of *Staphylococcus aureus*. *Int. J. Food Microbiol.* **2010**, *137*, 308–311. [CrossRef]
11. Hajibonabi, A.; Yekani, M.; Sharifi, S.; Nahad, J.S.; Dizaj, S.M.; Memar, M.Y. Antimicrobial activity of nanoformulations of carvacrol and thymol: New trend and applications. *OpenNano* **2023**, *13*, 100170. [CrossRef]
12. Mączka, W.; Twardawska, M.; Grabarczyk, M.; Wińska, K. Carvacrol—A natural phenolic compound with antimicrobial properties. *Antibiotics* **2023**, *12*, 824. [CrossRef]

13. Sánchez-Hernández, E.; Santiago-Aliste, A.; Correa-Guimarães, A.; Martín-Gil, J.; Gavara-Clemente, R.J.; Martín-Ramos, P. Carvacrol Encapsulation in Chitosan–Carboxymethylcellulose–Alginate Nanocarriers for Postharvest Tomato Protection. *Int. J. Mol. Sci.* **2024**, *25*, 1104. [CrossRef] [PubMed]
14. Laroque, D.A.; De Jong, N.R.; Müller, L.; Paganini, C.C.; de Araújo, P.H.; de Aragão, G.M.; Carciofi, B.A. Carvacrol release kinetics from cellulose acetate films and its antibacterial effect on the shelf life of cooked ham. *J. Food Eng.* **2023**, *358*, 111681. [CrossRef]
15. Shemesh, R.; Krepker, M.; Nitzan, N.; Vaxman, A.; Segal, E. Active packaging containing encapsulated carvacrol for control of postharvest decay. *Postharvest Biol. Technol.* **2016**, *118*, 175–182. [CrossRef]
16. Touil, H.F.; Boucherit, K.; Boucherit-Otmani, Z.; Kohder, G.; Madkour, M.; Soliman, S.S. Optimum inhibition of amphotericin-B-resistant *Candida albicans* strain in single- and mixed-species biofilms by *Candida* and non-*Candida terpenoids*. *Biomolecules* **2020**, *10*, 342. [CrossRef] [PubMed]
17. Yang, X.; Deng, P.; Liu, Q.; Meng, Y.; Dong, P.; Xu, L.; Huang, L. Exploring the efficacy of carvacrol as a biocontrol agent against pear Valsa canker. *Pestic. Biochem. Physiol.* **2023**, *196*, 105641. [CrossRef]
18. Leneveu-Jenvrin, C.; Aboudia, A.; Assemat, S.; Remize, F. A three-step approach to assess efficacy of alternative chemical treatments to preserve fresh fruit juices: Application to pineapple (*Ananas comosus* 'Queen Victoria'). *LWT-Food Sci. Technol.* **2022**, *155*, 112959. [CrossRef]
19. Churklam, W.; Chaturongakul, S.; Ngamwongsatit, B.; Aunpad, R. The mechanisms of action of carvacrol and its synergism with nisin against *Listeria monocytogenes* on sliced bologna sausage. *Food Control* **2020**, *108*, 106864. [CrossRef]
20. Peng, S.; Zhang, J.; Zhang, T.; Hati, S.; Mo, H.; Xu, D.; Li, H.; Hu, L.; Liu, Z. Characterization of carvacrol incorporated antimicrobial film based on agar/konjac glucomannan and its application in chicken preservation. *J. Food Eng.* **2022**, *330*, 111091. [CrossRef]
21. López-Córdoba, A. Feasibility of using carvacrol/starch edible coatings to improve the quality of paipa cheese. *Polymers* **2021**, *13*, 2516. [CrossRef] [PubMed]
22. Yang, X.; Fang, S.; Xie, Y.; Mei, J.; Xie, J. Preservative effects of flaxseed gum-sodium alginate active coatings containing carvacrol on quality of turbot (*Scophthalmus maximus*) during cold storage. *Coatings* **2024**, *14*, 338. [CrossRef]
23. Li, H.; Ding, J.; Liu, C.; Huang, P.; Yang, Y.; Jin, Z.; Qin, W. Carvacrol treatment reduces decay and maintains the postharvest quality of red grape fruits (*Vitis vinifera* L.) inoculated with Alternaria alternata. *Foods* **2023**, *12*, 4305. [CrossRef]
24. Zhao, L.; Wang, J.; Zhang, H.; Wang, P.; Wang, C.; Zhou, Y.; Li, H.; Yu, S.; Wu, R. Inhibitory effect of carvacrol against *Alternaria alternata* causing goji fruit rot by disrupting the integrity and composition of cell wall. *Front. Microbiol.* **2023**, *14*, 1139749. [CrossRef] [PubMed]
25. Duan, W.; Zhu, X.; Zhang, S.; Lv, Y.; Zhai, H.; Wei, S.; Ma, P.; Hu, Y. Antifungal effects of carvacrol, the main volatile compound in *Origanum vulgare* L. essential oil, against *Aspergillus flavus* in postharvest wheat. *Int. J. Food Microbiol.* **2024**, *410*, 110514. [CrossRef]
26. Ultee, A.; Bennik, M.; Moezelaar, R. The phenolic hydroxyl group of carvacrol is essential for action against the food-borne pathogen *Bacillus cereus*. *Appl. Environ. Microbiol.* **2002**, *68*, 1561–1568. [CrossRef]
27. Pei, S.; Liu, R.; Gao, H.; Chen, H.; Wu, W.; Fang, X.; Han, Y. Inhibitory effect and possible mechanism of carvacrol against *Colletotrichum fructicola*. *Postharvest Biol. Technol.* **2020**, *163*, 111126. [CrossRef]
28. Song, R.; Wang, X.; Jiao, L.; Jiang, H.; Yuan, S.; Zhang, L.; Shi, Z.; Fan, Z.; Meng, D. Epsilon-poly-l-lysine alleviates brown blotch disease of postharvest *Agaricus bisporus* mushrooms by directly inhibiting *Pseudomonas tolaasii* and inducing mushroom disease resistance. *Pestic. Biochem. Physiol.* **2024**, *199*, 105759. [CrossRef]
29. Zhang, Z.; Wang, P.; Chen, M.; Xie, L.; Zhang, X.; Shi, Y.; Lu, W.; Zhang, Q.; Li, C. Antibacterial activity of two new Cassane Diterpenoids from *Caesaplinia pulcherrima* against *Bacillus cereus* by damage to cell membrane. *Int. J. Mol. Sci.* **2023**, *24*, 4917. [CrossRef]
30. Ma, D.; Ji, D.; Liu, J.; Xu, Y.; Chen, T.; Tian, S. Efficacy of methyl thujate in inhibiting *Penicillium expansum* growth and possible mechanism involved. *Postharvest Biol. Technol.* **2020**, *161*, 111070. [CrossRef]
31. Bradford, M.M. A rapid and sensitive method for the quantitation of microgram quantities of protein utilizing the principle of protein-dye binding. *Anal. Biochem.* **1976**, *72*, 248–254. [CrossRef] [PubMed]
32. Morris, D.L. Quantitative determination of carbohydrates with Dreywood's anthrone reagent. *Science* **1948**, *107*, 254–255. [CrossRef] [PubMed]
33. Yang, X.; Yang, K.; Wang, X.; Wang, Y.; Zhao, Z.; Meng, D. Transcriptomic analysis reveals the mechanism of bacterial disease resistance of postharvest button mushroom (*Agaricus bisporus*). *Physiol. Mol. Plant Pathol.* **2022**, *122*, 101903. [CrossRef]
34. Nie, Z.; Huang, Q.; Chen, C.; Wan, C.; Chen, J. Chitosan coating alleviates postharvest juice sac granulation by mitigating ROS accumulation in harvested pummelo (*Citrus grandis* L. Osbeck) during room temperature storage. *Postharvest Biol. Technol.* **2020**, *169*, 111309. [CrossRef]
35. Selvakumar, P.; Rajasekar, S.; Periasamy, K.; Raaman, N. Isolation and characterization of melanin pigment from *Pleurotus cystidiosus* (telomorph of *Antromycopsis macrocarpa*). *World J. Microbiol. Biotechnol.* **2008**, *24*, 2125–2131. [CrossRef]
36. Sun, S.; Zhang, X.; Sun, S.; Zhang, L.; Shan, S.; Zhu, H. Production of natural melanin by *Auricularia auricula* and study on its molecular structure. *Food Chem.* **2016**, *190*, 801–807. [CrossRef]
37. Yang, R.R.; Han, Y.; Han, Z.H.; Ackah, S.; Li, Z.C.; Bi, Y.; Yang, Q.; Prusky, D. Hot water dipping stimulated wound healing of potato tubers. *Postharvest Biol. Technol.* **2020**, *167*, 111245. [CrossRef]

38. Ji, N.; Wang, J.; Li, Y.; Li, M.; Jin, P.; Zheng, Y. Involvement of *PpWRKY70* in the methyl jasmonate primed disease resistance against *Rhizopus stolonifer* of peaches via activating phenylpropanoid pathway. *Postharvest Biol. Technol.* **2021**, *174*, 111466. [CrossRef]
39. Wang, L.; Guo, Y.; Wang, X.; Zhang, X. Short-term O_2/CO_2 controlled atmosphere altered the water status and thus promoted phenolic biosynthesis during wound healing of fresh-cut white mushroom (*Agaricus bisporus*). *Postharvest Biol. Technol.* **2022**, *188*, 111879. [CrossRef]
40. Li, Z.; Xu, X.; Xue, S.; Gong, D.; Wang, B.; Zheng, X.; Xie, P.; Bi, Y.; Prusky, D. Preharvest multiple sprays with chitosan promotes the synthesis and deposition of lignin at wounds of harvested muskmelons. *Int. J. Biol. Macromol.* **2022**, *206*, 167–174. [CrossRef]
41. Jeon, M.; Zhao, Y. Honey in combination with vacuum impregnation to prevent enzymatic browning of fresh-cut apples. *Int. J. Food Sci. Nutr.* **2005**, *56*, 165–176. [CrossRef] [PubMed]
42. Arnold, M.; Gramza-Michałowska, A. Enzymatic browning in apple products and its inhibition treatments: A comprehensive review. *Compr. Rev. Food Sci. Food Saf.* **2022**, *21*, 5038–5076. [CrossRef] [PubMed]
43. Chung, I.-Y.; Kim, Y.-K.; Cho, Y.-H. Common virulence factors for *Pseudomonas tolaasii* pathogenesis in *Agaricus* and *Arabidopsis*. *Res. Microbiol.* **2014**, *165*, 102–109. [CrossRef]
44. Soler-Rivas, C.; Jolivet, S.; Arpin, N.; Olivier, J.; Wichers, H. Biochemical and physiological aspects of brown blotch disease of *Agaricus bisporus*. *FEMS Microbiol. Rev.* **1999**, *23*, 591–614. [CrossRef] [PubMed]
45. Dong, N.Q.; Lin, H. Contribution of phenylpropanoid metabolism to plant development and plant–environment interactions. *J. Integr. Plant Biol.* **2021**, *63*, 180–209. [CrossRef]
46. Dawadi, E.; Magar, P.; Bhandari, S.; Subedi, S.; Shrestha, S.; Shrestha, J. Nutritional and post-harvest quality preservation of mushrooms: A review. *Heliyon* **2022**, *8*, e12093. [CrossRef]
47. Cao, Z.; Zhou, D.; Ge, X.; Luo, Y.; Su, J. The role of essential oils in maintaining the postharvest quality and preservation of peach and other fruits. *J. Food Biochem.* **2022**, *46*, e14513. [CrossRef]
48. Mechmechani, S.; Gharsallaoui, A.; Fadel, A.; El Omari, K.; Khelissa, S.; Hamze, M.; Chihib, N.-E. Microencapsulation of carvacrol as an efficient tool to fight *Pseudomonas aeruginosa* and *Enterococcus faecalis* biofilms. *PLoS ONE* **2022**, *17*, e0270200. [CrossRef] [PubMed]
49. Kasthuri, T.; Swetha, T.K.; Bhaskar, J.P.; Pandian, S.K. Rapid-killing efficacy substantiates the antiseptic property of the synergistic combination of carvacrol and nerol against nosocomial pathogens. *Arch. Microbiol.* **2022**, *204*, 590. [CrossRef]
50. Zhang, Z.; Tan, Y.; McClements, D.J. Investigate the adverse effects of foliarly applied antimicrobial nanoemulsion (carvacrol) on spinach. *LWT-Food Sci. Technol.* **2021**, *141*, 110936. [CrossRef]
51. Xiao, L.; Kang, S.; Lapu, M.; Jiang, P.; Wang, X.; Liu, D.; Li, J.; Liu, M. Preparation and characterization of chitosan/pullulan film loading carvacrol for targeted antibacterial packaging of chilled meat. *Int. J. Biol. Macromol.* **2022**, *211*, 140–149. [CrossRef] [PubMed]
52. Fonseca, L.M.; Souza, E.J.D.; Radünz, M.; Gandra, E.A.; da Rosa Zavareze, E.; Dias, A.R.G. Suitability of starch/carvacrol nanofibers as biopreservatives for minimizing the fungal spoilage of bread. *Carbohydr. Polym.* **2021**, *252*, 117166. [CrossRef]
53. Mauriello, E.; Ferrari, G.; Donsì, F. Effect of formulation on properties, stability, carvacrol release and antimicrobial activity of carvacrol emulsions. *Colloids Surf. B* **2021**, *197*, 111424. [CrossRef] [PubMed]
54. Fang, J.; Yin, Z.; Zhang, T.; Yang, W.; Fang, T.; Wang, Y.; Guo, N. Preparation and characterization of carvacrol/ε-polylysine loaded antimicrobial nanobilayer emulsion and its application in mango preservation. *Food Chem.* **2024**, *446*, 138831. [CrossRef] [PubMed]
55. Jiang, S.; Zhang, J.; Yang, Q.; Sun, D.; Pu, X.; Shen, H.; Li, Q.; Wang, Z.; Lin, B. Antimicrobial activity of natural plant compound carvacrol against soft rot disease agent *Dickeya zeae*. *Curr. Microbiol.* **2021**, *78*, 3453–3463. [CrossRef] [PubMed]
56. Yin, J.; Wu, S.; Yang, Y.; Wang, D.; Ma, Y.; Zhao, Y.; Sheth, S.; Huang, H.; Song, B.; Chen, Z. In addition to damaging the plasma membrane, phenolic monoterpenoid carvacrol can bind to the minor groove of DNA of phytopathogenic fungi to potentially control tea leaf spot caused by *Lasiodiplodia theobromae*. *Phytopathology* **2024**, *114*, 700–716. [CrossRef]
57. Chaillot, J.; Tebbji, F.; Remmal, A.; Boone, C.; Brown, G.W.; Bellaoui, M.; Sellam, A. The monoterpene carvacrol generates endoplasmic reticulum stress in the pathogenic fungus *Candida albicans*. *Antimicrob. Agents Chemother.* **2015**, *59*, 4584–4592. [CrossRef]
58. Wang, L.; Guo, Y.; Wang, X. Increasing nicotinamide adenine dinucleotide phosphate hydrogen levels via pentose phosphate pathway promotes phyenylpropanoid metabolism of fresh-cut white mushroom (*Agaricus bisporus*) under high O_2/CO_2 storage. *Sci. Hortic.* **2022**, *295*, 110873. [CrossRef]
59. Zhang, P.; Fang, D.; Pei, F.; Wang, C.; Jiang, W.; Hu, Q.; Ma, N. Nanocomposite packaging materials delay the browning of *Agaricus bisporus* by modulating the melanin pathway. *Postharvest Biol. Technol.* **2022**, *192*, 112014. [CrossRef]
60. Du, M.; Lian, L.; Zhang, Y.; Lin, H.; Wang, J. Roles of ROS metabolism and phenylpropanoid pathway in quality maintenance of postharvest *Pleurotus eryngii* under hyperoxia stress. *Postharvest Biol. Technol.* **2024**, *207*, 112617. [CrossRef]
61. Singh, D.; Mittal, N.; Mittal, P.; Siddiqui, M.H. Transcriptome sequencing of medical herb *Salvia rosmarinus* (Rosemary) revealed the phenylpropanoid biosynthesis pathway genes and their phylogenetic relationships. *Mol. Biol. Rep.* **2024**, *51*, 757. [CrossRef] [PubMed]
62. Mi, T.; Luo, D.; Li, J.; Qu, G.; Sun, Y.; Cao, S. Carvacrol exhibits direct antifungal activity against stem-end rot disease and induces disease resistance to stem-end rot disease in kiwifruit. *Physiol. Mol. Plant Pathol.* **2023**, *127*, 102065. [CrossRef]

63. Yan, J.; Yuan, S.-S.; Jiang, L.-L.; Ye, X.-J.; Ng, T.B.; Wu, Z.-J. Plant antifungal proteins and their applications in agriculture. *Appl. Microbiol. Biotechnol.* **2015**, *99*, 4961–4981. [CrossRef]
64. Zhao, Y.; Li, C.; Chen, X.; Cheng, J.; Xie, J.; Lin, Y.; Fu, Y.; Jiang, D.; Chen, T. Overexpression of chitinase *PbChia1* from *Plasmodiophora brassicae* improves broad-spectrum disease resistance of *Arabidopsis*. *Virulence* **2023**, *14*, 2233147. [CrossRef] [PubMed]

foods
MDPI

Article

Effects of Consuming Pulsed UV Light-Treated *Pleurotus citrinopileatus* on Vitamin D Nutritional Status in Healthy Adults

Chih-Ching Hsu [1,†], Chiao-Ming Chen [2,†], Yu-Ming Ju [3], Yu-Ching Wu [3], Huei-Mei Hsieh [3], Shu-Hui Yang [4], Chien-Tien Su [5,6], Te-Chao Fang [7,8,9], Widiastuti Setyaningsih [10] and Sing-Chung Li [1,*]

1 School of Nutrition and Health Sciences, College of Nutrition, Taipei Medical University, Taipei 11031, Taiwan; ga56110007@tmu.edu.tw
2 Department of Food Science, Nutrition, and Nutraceutical Biotechnology, Shih Chien University, Taipei 10462, Taiwan; charming@g2.usc.edu.tw
3 Institute of Plant and Microbial Biology, Academia Sinica, Taipei 11529, Taiwan; yumingju@gate.sinica.edu.tw (Y.-M.J.); saseki@gate.sinica.edu.tw (Y.-C.W.); monicah@gate.sinica.edu.tw (H.-M.H.)
4 Taiwan Agricultural Research Institute, Fengshan Tropical Horticultural Experiment Branch, Kaohsiung City 83052, Taiwan; debbie@tari.gov.tw
5 Department of Family Medicine, Taipei Medical University Hospital, Taipei 11031, Taiwan; ctsu@tmu.edu.tw
6 School of Public Health, College of Public Health, Taipei Medical University, Taipei 11031, Taiwan
7 Division of Nephrology, Department of Internal Medicine, School of Medicine, College of Medicine, Taipei Medical University, Taipei 11031, Taiwan; fangtc@tmu.edu.tw
8 Division of Nephrology, Department of Internal Medicine, Taipei Medical University Hospital, Taipei Medical University, Taipei 11031, Taiwan
9 Taipei Medical University-Research Center of Urology and Kidney, Taipei Medical University, Taipei 11031, Taiwan
10 Department of Food and Agricultural Product Technology, Faculty of Agricultural Technology, Universitas Gadjah Mada, Jalan Flora, Depok, Sleman, Yogyakarta 55281, Indonesia; widiastuti.setyaningsih@ugm.ac.id
* Correspondence: sinchung@tmu.edu.tw; Tel.: +886-2-27361661 (ext. 6560)
† These authors contributed equally to this work.

Abstract: Vitamin D, essential for growth and health, is often deficient in Taiwan despite abundant sunlight. Plant-derived vitamin D_2 (ergocalciferol) is bioavailable, environmentally friendly, and cost-effective. This study evaluated the efficacy of enhancing *Pleurotus citrinopileatus* (PC) mushrooms' vitamin D_2 content through pulsed ultraviolet (PUV) light and its impact on vitamin D status in humans. In a four-week randomized parallel trial, 36 healthy participants were assigned to three groups: a control group, a group consuming 10 g/day PUV-treated PC (PC-10 g), and a group consuming 100 g/day PUV-treated PC (PC-100 g). Blood samples collected pre- and post-intervention measured serum 25(OH)D_2, 25(OH)D_3, and biochemical parameters. After four weeks, serum 25(OH)D_2 levels significantly increased in the PC-10 g group (1.47 ± 1.42 ng/mL to 9.50 ± 7.10 ng/mL, $p = 0.001$) and in the PC-100 g group (1.94 ± 2.15 ng/mL to 21.82 ± 16.75 ng/mL, $p = 0.002$), showing a 10.2-fold rise. The PC-100 g group also experienced a 37.6% reduction in serum intact parathyroid hormone (I-PTH) levels (26.26 ± 9.84 pg/mL to 16.38 ± 5.53 pg/mL). No adverse effects were reported. PUV-treated PC mushrooms significantly increase serum 25(OH)D_2 levels and reduce I-PTH, particularly at higher doses. These findings underscore the potential of vitamin-D-enriched PC as a sustainable, fungi-derived food source for addressing vitamin D deficiency.

Keywords: vitamin D_2; ergosterol; *Pleurotus citrinopileatus*; pulsed ultraviolet light; 25(OH)D_2; intact parathyroid hormone

https://doi.org/10.3390/foods14020259

1. Introduction

Vitamin D, a steroid hormone, is a vital nutrient for humans, playing crucial roles in bone growth, calcium and phosphorus balance, and immune regulation. Vitamin D is classified into two types based on its source: vitamin D_2 (ergocalciferol) and vitamin D_3 (cholecalciferol). Vitamin D_2 primarily comes from fungi-derived sources, while vitamin D_3 is derived from animal-based sources or synthesized in the skin through sunlight exposure. Ergosterol, a fungi-derived precursor of vitamin D_2, can be converted into ergocalciferol through ultraviolet light activation and subsequently metabolized into its active form for utilization in the body [1–3].

Although Taiwan is in the subtropical zone and generally experiences ample sunlight, a high proportion of the population still suffers from vitamin D deficiency. Lee et al. reported that among participants aged $\geq$ 30 years without chronic kidney disease (CKD) in northern Taiwan, 22.4% of the study population had vitamin D deficiency (25(OH)D level < 20 ng/mL). Additionally, the study observed a significantly higher prevalence of vitamin D deficiency in women compared to men [4]. Due to the health and environmental benefits associated with vegetarianism, the vegetarian population in Taiwan has been increasing. Several studies have shown that vegetarians and vegans exhibit the highest prevalence of vitamin D deficiency [5,6]. Similarly, a small study in Taiwan reported that individuals adhering to a strict vegetarian diet had significantly lower vitamin D levels compared to non-vegetarians [7].

Mushrooms naturally contain high concentrations of ergosterol in their cell walls. Upon exposure to ultraviolet (UV) irradiation, ergosterol is converted into previtamin D_2, which subsequently undergoes a temperature-dependent thermal isomerization process to form ergocalciferol, also known as vitamin D_2 [8]. However, most fresh retail mushrooms are cultivated in atmospherically controlled growing rooms under dark conditions, resulting in a vitamin D_2 content typically less than 1 μg per 100 g of fresh weight (FW) [9]. To produce nutritionally relevant levels of vitamin D_2, mushrooms can be exposed to controlled UV irradiation using fluorescent or pulsed UV lamps. This exposure can be applied during their growth stage, after harvest, or following the drying process [10].

Since vitamin D_2 is primarily found in fungi and yeast, common fruits and vegetables, which lack ergosterol, cannot serve as a source of vitamin D_2 intake [11]. *Pleurotus citrinopileatus* (PC) is a nutritionally valuable edible fungus rich in bioactive compounds, including proteins, polysaccharides, amino acids, minerals, dietary fiber, and trace elements. It is widely consumed in Taiwan [12] and cultivated globally, with East Asia and northeastern China as major production regions [13]. Huang et al. investigated vitamin D_2 production in various mushroom species exposed to UVB irradiation and found that PC exhibited the highest conversion efficiency. In PC, the vitamin D_2 content increased significantly from 3.93 ± 0.44 μg/g to 208.65 ± 6.08 μg/g dry matter after 2 h of UVB exposure, representing an approximate 53-fold increase [14].

Several studies have explored the effects of UVB irradiation on mushroom quality. While UVB exposure enhances nutritional components such as phenolics, flavonoids, and vitamin D_2, it also leads to quality deterioration, including darkening, texture changes, and moisture exudation. These issues adversely affect the appearance, marketability, and shelf-life of mushrooms [15–17]. In contrast, Kalaras and Koyyalamudi studied the effects of postharvest pulsed ultraviolet (PUV) light on white button mushrooms (*Agaricus bisporus*), finding that PUV treatment significantly boosted vitamin D_2 levels without compromising appearance, texture, or shelf-life, making it an effective method for enhancing nutritional value while preserving quality [10,18].

This study explores the effectiveness of PUV light in enhancing the vitamin D_2 content in PC mushrooms and its impact on healthy adult vitamin D status. It evaluates whether

PUV can increase vitamin D_2 levels and improve nutritional status after four weeks of intake and whether the rise in blood vitamin D levels is attributed to increased 25(OH)D_2 concentrations, highlighting the novelty of this research.

2. Materials and Methods

2.1. PUV Light Treatment PC

The fresh PC mushrooms used in this experiment were cultivated by a professional mushroom farmer in Yuchi Township, Nantou County. The PUV light treatment was performed by Dr. Yang from the Kaohsiung-Fengshan Tropical Horticulture Experiment Office. Freshly harvested PC mushrooms were exposed to PUV light using a Xenon Z-1000-OEM device (MOS Technology Inc., Hsinchu, Taiwan). The equipment had an energy specification of 1516 J/s with a flash frequency of 3 Hz. The irradiated area was 30 cm × 8 cm = 240 cm^2, resulting in an energy density of 6.31 J/s·cm^2. Each square centimeter of the mushrooms was exposed for 1.5 s, delivering a total energy of 9.465 J/cm^2. This treatment was optimized to enhance the vitamin D_2 content of the mushrooms while preserving their quality.

2.2. Preparation of PC Instant Meal Packs for Participants

PC mushrooms treated with PUV light were transported under refrigeration to the experimental kitchen at Shih Chien University. After cleaning, the mushrooms were steamed for 20 min, seasoned with vegetable oil and spices, vacuum-sealed, and frozen at −20 °C. The preparation process was carried out weekly to ensure quality, with PC instant meal packs prepared in two portions: 10 g and 100 g. Frozen mushroom meal packs were distributed to participants in weekly portions, with instructions to heat them using a water bath before consumption. Participants were instructed to consume one pack daily with a meal.

2.3. Extraction and Analysis of Vitamin D_2 in PC Mushrooms

Vitamin D_2 was extracted and analyzed following the method of Huang et al. (2015) with modifications [14]. The vitamin D_2 content in the PC exposed to PUV light was measured at two stages: before cooking and after cooking. The process began by freeze-drying the PC, followed by weighing 5 g of the powdered sample into a centrifuge tube. Next, 10 mL of dimethyl sulfoxide (DMSO) was added, and the mixture was sonicated in a water bath at 40–50 °C for 30 min. Subsequently, 10 mL of a methanol solution (1:1, v/v) was added and thoroughly mixed. Afterward, 20 mL of n-hexane was introduced, and the mixture was agitated for 30 min. The sample was centrifuged at 3500 rpm for 10 min, and the upper layer was collected into a concentration flask. The lower layer was re-extracted twice more with 20 mL of n-hexane. The combined upper layers were concentrated to near dryness under reduced pressure in a water bath at 30–40 °C. The residue was dissolved in n-hexane and diluted to 5 mL. The final solution was filtered through a membrane filter for analysis.

Vitamin D_2 analysis in the PC mushrooms was performed using a Waters® ACQUITY UPLC system (Water, Milford, CT, USA) equipped with a Vydac C18 column (5 μm, 4.6 × 250 mm). The detection wavelength was set at 265 nm, with a mobile phase of 10:90 (v/v) methanol/acetonitrile at a flow rate of 0.5 mL/min. The retention time for vitamin D_2 was 2.1 min.

2.4. Participant Recruitment

Thirty-six healthy participants were recruited at Taipei Medical University, Shih Chien University, and the Family Medicine Health Check Center at Taipei Medical University

Hospital. Inclusion criteria were as follows: aged 20 to 60 years, any gender, and able to consume one mushroom meal pack daily. Exclusion criteria included the following: (1) Individuals with any acute diseases such as infections, stroke, myocardial infarction, major surgery within the past three months, upper or lower gastrointestinal bleeding, and poorly controlled blood pressure or blood glucose. (2) Individuals with conditions such as malignant tumors, human immunodeficiency virus (HIV), cirrhosis or liver function exceeding three times the normal value (AST or ALT > 120 IU/L), abnormal kidney function (creatinine greater than 2.5 mg/dL), anemia (Hb < 9 g/dL), metabolic disorders other than diabetes (e.g., thyroid or parathyroid dysfunction), or previous abdominal surgery resulting in intestinal adhesions. (3) Pregnant or breastfeeding women. (4) Individuals taking steroids, vitamin D supplements, or hormones. (5) Participants who had had a previous allergy to mushrooms. (6) Participants with serum 25(OH)D levels > 30 ng/mL.

This study was approved by the Taipei Medical University Institutional Review Board (N202111054). This study has also been registered with ClinicalTrials.gov, with the registration number NCT05716698. The study was initiated after obtaining informed consent from participants who fully understood the study's objectives, procedures, and the guidelines to be followed during the study.

2.5. Clinical Trial Design

Thirty-six healthy participants were randomly assigned to one of three groups in a parallel design: (1) a control group with no PC intervention (Control), (2) a group consuming 10 g of PUV-treated PC daily (PC-10 g), and (3) a group consuming 100 g of PUV-treated PC daily (PC-100 g). Each group consisted of 12 participants and underwent a four-week experimental intervention.

Prior to trial entry, participants who had been regularly taking any form of vitamin D supplements were excluded. Although a specific washout period was not implemented, baseline serum 25(OH)D levels were measured, and only individuals with levels below 30 ng/mL were included in the study, ensuring that all participants were either vitamin-D-insufficient or -deficient at the start of the trial.

Before and after the trial, participants were required to fast for at least eight hours and then undergo venous blood collection of 11 mL for biochemical testing and analysis of serum 25(OH)D_2 and 25(OH)D_3. During the intervention period, participants were reminded daily via the LINE messaging app to consume the PC meal packs and to report any adverse effects. Participants visited the laboratory weekly to pick up their PC meal packs and return any empty or uneaten packs, which helped to ensure compliance. We ensured that the compliance rate among the participants was nearly 100%, and none of the participants reported any adverse side effects. All participants maintained the same general activities and regular diet during the trial as before the trial.

2.6. Blood Biochemical Measurements

Blood biochemical analyses, including fasting plasma glucose, triglycerides, total cholesterol, low-density lipoprotein cholesterol (LDL-C), high-density lipoprotein cholesterol (HDL-C), aspartate aminotransferase (AST), alanine aminotransferase (ALT), blood urea nitrogen (BUN), creatinine, calcium, phosphorus, magnesium, high-sensitivity C-reactive protein (hs-CRP), glycated hemoglobin (HbA1c), and intact parathyroid hormone (iPTH), were performed using standardized methods at Yadong Medical Laboratory (Zhongli, Taiwan). Specifically, fasting plasma glucose was measured using a Beckman Coulter AU5800 automated chemistry analyzer (Beckman Coulter, Brea, CA, USA), and HbA1c was analyzed with a Premier Hb9210™ automated analyzer (Trinity Biotech, Wicklow, Ireland).

2.7. Serum 25(OH)D_2 and 25(OH)D_3 Analysis

During the study, 4 mL of whole blood was collected from the 36 participants at baseline and on day 28, and serum was separated for the measurement of 25(OH)D_2 and 25(OH)D_3, and the extraction preparation of blood samples followed the method by Chen et al. [19].

First, 0.5 mL of serum was mixed with 2 mL of ethanol and shaken for 15 s. Then, 2 mL of deionized water was added, and the mixture was shaken for another 15 s. Next, 3 mL of hexane was added and shaken for 1 min. The mixture was then centrifuged using a HITACHI Centrifuge CT6E (HITACHI, Japan) at 10,000 rpm for 10 min to collect the upper organic layer. The extraction process was repeated with an additional 3 mL of hexane, and the supernatant was collected. The two extracts were combined and freeze-dried using a Scanvac ScanSpeed 32 Teflon freeze-dryer. Prior to analysis, 100 µL of methanol was added to reconstitute the residue.

This study utilized the ultra-high-performance liquid chromatography–mass spectrometry (UHPLC-MS) method of analysis. The detection of 25(OH)D_2 and 25(OH)D_3 was performed using a Thermo Scientific Vanquish Horizon UHPLC system coupled with atmospheric pressure chemical ionization (APCI) and an ion trap analyzer, specifically, an Orbitrap Fusion Lumos tribrid mass spectrometer (Thermo Scientific). The UHPLC injection temperature was set at 5 °C, and the column used was an ACQUITY UPLC CSH C18, 1.7 µm (2.1 × 50 mm, Waters). The column temperature was set to 40 °C for separation. Mobile phase A was 25% acetonitrile (ACN), and mobile phase B was a mixture of ACN and methanol in a 3:1 ratio. The flow rate was set to 0.4 mL/min. Initially, the mobile phase ratio was 20% A and 80% B, which was maintained for 4 min. Then, the proportion of mobile phase B was increased to 100% and maintained for 1 min. At 5.1 min, the proportion of mobile phase B was decreased to 80%, and this ratio was maintained until the end of the run at 6.5 min. High-energy collision-induced dissociation was employed for the analysis. The ionization temperature was set at 350 °C, and the ion transfer temperature was set at 275 °C. The spray voltage was 4 µA in the positive mode, and the sheath gas flow rate was set to 45 arbitrary units. The monitoring time was 8 min.

2.8. Statistical Methods

All data are presented as mean ± standard deviation (SD). The Shapiro–Wilk test was used to assess the normality of the data distribution. Within-group comparisons before and after the intervention were conducted using paired *t*-tests. Differences between groups were analyzed using one-way analysis of variance (ANOVA) followed by Tukey's post hoc Honestly Significant Difference (HSD) test. Statistical significance was set at $p < 0.05$. All analyses were performed using IBM SPSS Statistics, Version 19.

3. Results

3.1. Vitamin D_2 Content in PUV-Light-Treated PC

The non-PUV-treated fresh PC contained extremely low levels of vitamin D_2, all below 1 µg per 100 g of fresh weight, which is under the detection limit of UPLC. After PUV treatment, the PC had an average vitamin D_2 content of 248 µg per 100 g. Following cooking, approximately 7.8% of the vitamin D_2 was lost. Therefore, the mushroom meal packs provided to the participants contained an average of 248 µg of vitamin D_2 per 100 g (Table 1).

Table 1. Vitamin D_2 content in Pleurotus citrinopileatus exposed to PUV across different batches.

Batches	Fresh PC Mushroom (µg /100 g)	Cooked PC Mushroom (µg /100 g)	Cooking Loss Rate (%)
First Batch	218.1 ± 0.33	199.7 ± 0.75	8.5
Second Batch	278.3 ± 11.6	252.1 ± 14.1	9.4
Third Batch	250.9 ± 17.8	235.2 ± 5.9	6.3
Fourth Batch	245.4 ± 10.1	228.6 ± 1.5	6.9
Average	248.2 ± 10.0	228.9 ± 5.6	7.8

Data in the table are expressed as mean ± SD.

3.2. Subjects Recruitment and Basic Characteristics of the Participants

Recruitment for this study began in March 2022. By April 2022, a total of 42 participants were recruited. However, three participants withdrew before the experiment began due to concerns about the COVID-19 pandemic, one participant withdrew before starting the study due to health issues, and two participants withdrew without providing a reason. Ultimately, 36 participants (5 males and 31 females) met the inclusion criteria, signed the consent forms, and completed the study (Figure 1).

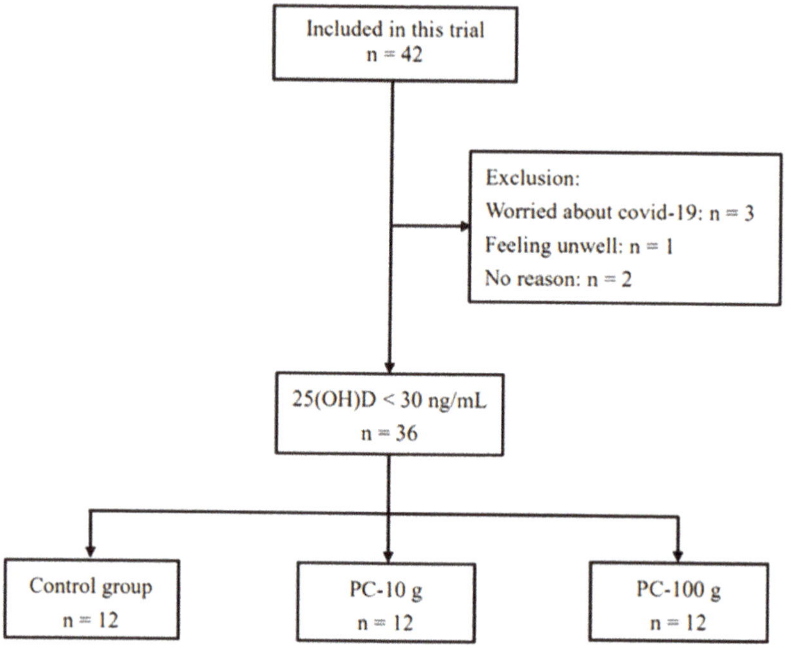

Figure 1. Flowchart of subject recruitment.

The basic characteristics of the participants are presented in Table 2. The results indicate that the control group (31.3 ± 12.9 yrs) had a significantly higher mean age compared to the PC-10 g group (21.9 ± 2.0 yrs), while the PC-100 g group (24.3 ± 5.7 yrs) showed no significant difference from the other two groups. There were no other statistical differences between the groups in terms of gender distribution, education level, or medical history.

Table 2. Blood biochemical measurements of the healthy subjects.

Characteristic	Control (*N* = 12)	PC-10 g (*N* = 12)	PC-100 g (*N* = 12)	*p*
Age (years)	31.3 ± 12.9	21.9 ± 2.0	24.3 ± 5.7	0.024 *
Male/female (n)	2/10	1/11	2/10	0.807
Education				
High school (n)	1	0	0	
College (n)	11	12	12	
Medical history				
Hypertension (n)	1	0	0	
Hyperlipidemia (n)	1	1	2	
Diabetes (n)	0	0	0	

Data are expressed as mean ± standard deviation or numbers and analyzed by one-way ANOVA followed by Tukey's post hoc Honestly Significant Difference (HSD) test. $p < 0.05$ was considered as statistically significant and is shown with an asterisk.

3.3. Blood Biochemical Measurements of the Participants

This study randomly assigned 36 subjects into three groups: the control group (n = 12), the PC-10 g group (n = 12), and the PC-100 g group (n = 12). Blood biochemical parameters, including fasting plasma glucose (FPG), total cholesterol, triglycerides (TG), aspartate transaminase (AST), alanine transaminase (ALT), blood urea nitrogen (BUN), creatinine, high-sensitivity C-reactive protein (Hs-CRP), magnesium (Mg), calcium (Ca), and phosphorus (P), were measured at baseline (day 0) and after 28 days of intervention. The results, as shown in Table 3, demonstrated no significant changes in any parameters in the control group ($p > 0.05$). Similarly, in both the PC-10 g and PC-100 g groups, where the participants consumed 10 g and 100 g of PUV-treated PC daily, none of the measured biochemical parameters showed statistically significant differences between day 0 and day 28 ($p > 0.05$). Furthermore, in the PC-10 and PC-100 g group, which received a high dose of vitamin D_2, no adverse effects were observed in the liver (AST, ALT) or kidney (BUN, creatinine) indicators, suggesting that the consumption of the UV-treated PC was safe and did not affect the blood biochemical parameters in the subjects.

Table 3. Blood biochemical measurements of the subjects.

Items	Control (*N* = 12)			PC-10 g (*N* = 12)			PC-100 g (*N* = 12)		
	Day 0	Day 28	*p*	Day 0	Day 28	*p*	Day 0	Day 28	*p*
FPG (mg/dL)	81.8 ± 5.8	81.7 ± 4.1	0.936	84.3 ± 5.3	82.4 ± 6.2	0.424	81.7 ± 4.7	79.7 ± 7.4	0.436
Cholesterol (mg/dL)	187.8 ± 40.7	179.2 ± 28.9	0.553	172.1 ± 32.7	180.3 ± 36.7	0.567	200.1 ± 35.3	198.6 ± 38.8	0.922
TG (mg/dL)	85.6 ± 54.5	89.3 ± 57.0	0.874	98.3 ± 110.9	87.2 ± 81.4	0.783	83.3 ± 46.2	77.3 ± 21.8	0.688
AST (U/L)	17.8 ± 5.2	20.2 ± 5.5	0.294	24.3 ± 22.9	19.3 ± 5.9	0.464	17.8 ± 5.1	20.2 ± 6.1	0.321
ALT (U/L)	13.7 ± 5.1	14.4 ± 5.7	0.736	15.9 ± 12.06	16.4 ± 15.61	0.931	17.2 ± 17.1	19.5 ± 19.7	0.759
BUN (mg/dL)	14.4 ± 3.2	13.7 ± 2.9	0.554	11.3 ± 2.0	11.1 ± 3.1	0.877	11.7 ± 2.1	12.5 ± 2.4	0.370
Creatinine (mg/dL)	0.7 ± 0.1	0.7 ± 0.1	0.611	0.7 ± 0.1	0.6 ± 0.1	0.352	0.7 ± 0.1	0.7 ± 0.1	0.442
Hs-CRP (mg/L)	0.57 ± 0.57	1.27 ± 2.17	0.290	0.36 ± 0.30	0.48 ± 0.38	0.387	0.74 ± 0.96	0.52 ± 0.53	0.481
Mg (mg/dL)	2.07 ± 0.14	1.97 ± 0.11	0.059	1.99 ± 0.14	1.98 ± 0.17	0.898	2.03 ± 0.16	1.91 ± 0.17	0.095
Ca (mg/dL)	9.42 ± 0.32	9.18 ± 0.34	0.093	9.40 ± 0.31	9.35 ± 0.23	0.655	9.42 ± 0.26	9.40 ±0.26	0.878
P (mg/dL)	3.98 ± 0.41	3.70 ± 0.55	0.166	4.15 ± 0.49	3.87 ± 0.38	0.126	3.94 ± 0.38	3.88 ± 0.29	0.679

Data are expressed as mean ± standard deviation and were analyzed by paired *t*-test; FPG, fasting plasma glucose; TG, triglycerides; AST, aspartate transaminase; ALT, alanine transaminase; BUN, blood urea nitrogen; Hs-CRP, high-sensitivity C-reactive protein. $p < 0.05$ was considered as statistically significant.

3.4. Serum 25(OH)D_2 and 25(OH)D_3 Levels of Participants Before and After the Trial

The changes in serum 25(OH)D levels before and after the trial are shown in Figure 2. The results showed that before PC supplementation, the serum 25(OH)D_2 levels of all subjects were very low. After the intervention, except for the no differences observed in the control group before and after intervention, the PC-10 g group showed a significant increase in serum 25(OH)D_2 from 1.47 ± 1.42 ng/mL to 9.50 ± 7.10 ng/mL after four weeks ($p < 0.001$), while that of the PC-100 g group increased significantly from 1.94 ± 2.15 ng/mL to 21.82 ± 16.75 ng/mL ($p < 0.001$). The serum 25(OH)D_2 levels significantly increased with the dosage of the supplement. However, there was no significant difference in serum 25(OH)D_3 levels before and after supplementation. This indicates that consuming PUV-treated PC can significantly enhance the vitamin D nutritional status of healthy subjects.

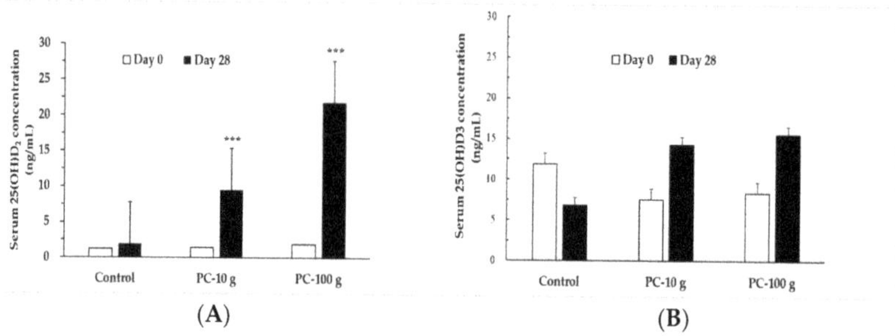

Figure 2. Serum 25(OH)D_2 (**A**) and 25(OH)D_3 (**B**) concentration in subjects before and after PC intervention, measured by LC-MS analysis. *** means $p < 0.001$.

3.5. Regulatory Effect of High-Dose PUV-Treated PC Supplementation on Serum Parathyroid Hormone Levels

A decrease in 25(OH)D reduces calcium absorption, leading to lower serum calcium levels, which triggers the release of parathyroid hormone (PTH). Notably, high-dose PC supplementation significantly increased serum 25(OH)D_2 levels and decreased serum intact PTH levels by 37.6% ($p = 0.013$). This finding suggests that high-dose PC may improve PTH levels, likely due to the enhanced intake of vitamin D_2 (Figure 3).

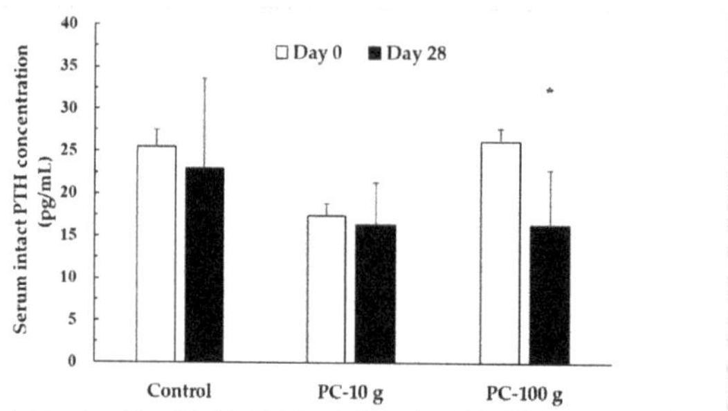

Figure 3. Serum intact PTH concentration in subjects before and after PC intervention. PTH, parathyroid hormone. * $p > 0.001$.

4. Discussion

Vitamin D plays a vital role in the body, including enhancing the intestinal absorption of calcium, magnesium, and phosphorus. Vitamin D deficiency has become a global concern and is associated with various diseases, including bone disorders. However, large-scale vitamin D supplementation trials have yielded mixed results, highlighting the need for further research to clarify its specific effects [20]. A meta-analysis by Cui et al., incorporating data from 7.9 million participants across 81 countries, revealed that between 2000 and 2022, the global prevalence of vitamin D levels below 75 nmol/L was approximately 76.6%. The study found that Africa had the lowest prevalence, while the eastern Mediterranean region exhibited the highest. Higher prevalence rates were observed among females, during winter and spring seasons, in high-latitude regions, and in low- to middle-income countries [21]. According to the Korea National Health and Nutrition Examination Survey (KNHANES) data from 2016 to 2019, the average dietary intake of vitamin D among Koreans is significantly lower than the recommended daily intake, averaging only 3.13 micrograms per day. Further analysis of dietary sources reveals that fish and shellfish are the primary contributors, accounting for 61.59%, followed by eggs (17.75%), meat (8.03%), milk (4.25%), legumes (3.93%), and grains (3.84%). This survey also highlights that animal-based sources remain the predominant means of obtaining vitamin D for Koreans [22]. Lee et al. also found similar results in Taiwan [23]. Jiang et al. found that 83% of Chinese adults aged 18 to 65 had serum 25(OH)D_3 levels below 30 ng/mL, with 41.9% showing deficiency (10–19 ng/mL) and 8.4% severe deficiency (<10 ng/mL). Women, winter and spring seasons, northern regions, and younger age groups were associated with lower vitamin D levels [24]. The most recent National Nutrition and Health Survey in Taiwan (NAHSIT, 2017–2020) indicated that adult males consumed only 52% to 57% of the adequate intake (AI) for vitamin D recommended in the eighth edition of the Dietary Reference Intakes (DRIs), while females consumed just 38% to 49%. This inadequate intake was reflected in the serum 25(OH)D_2 levels of the participants in our study, which were consistently found to be extremely low before the intervention.

Fungi-derived sources of vitamin D offer advantages such as environmental friendliness and lower costs, which may contribute to improving human vitamin D nutritional status. Wild mushrooms can naturally absorb UV light, typically containing vitamin D_2. However, modern farming practices, which focus on producing high-quality mushrooms, often involve growing mushrooms under controlled light, humidity, and temperature conditions. As a result, vitamin D_2 levels in these cultivated mushrooms are usually undetectable [11]. Traditional UV irradiation methods are effective in producing high concentrations of vitamin D_2 and have also been shown to enhance antioxidant capacity [25]; however, they often lead to issues such as blackening, exudation, and reduced shelf-life of fresh mushrooms due to prolonged UVB exposure. In contrast, the pulsed UV light (PUV) method described by Kalaras et al. does not affect the color or quality of fresh mushrooms, making it an optimal approach for producing mushrooms with high concentrations of vitamin D_2 [10,26].

Several studies have demonstrated that vitamin D_2 in mushrooms is bioavailable. Urbain et al. conducted a single-blind, randomized, placebo-controlled trial over five weeks with 26 healthy adults whose 25(OH)D levels were below 20 ng/mL. The participants were randomly assigned to one of three groups: one receiving UV-irradiated mushroom soup (providing 28,000 IU per week), one receiving the same dose of vitamin D_2 supplements, and a placebo group (non-UV-irradiated mushroom soup). At the end of the trial, the serum 25(OH)D levels significantly increased in the UV-irradiated mushroom soup group, showing similar effectiveness to the vitamin D_2 supplements in raising 25(OH)D levels [27]. Stephensen et al. found that after providing participants with UV-untreated mushroom meals (control), UV-irradiated mushroom meals containing 352 IU or 684 IU of vitamin D_2, and purified ergocalciferol (1128 IU) plus untreated mushrooms for six weeks, all groups except the control group showed increased levels of 25(OH)D_2. However, there was a decrease in serum 25(OH)D_3 levels across all groups consuming vitamin D_2. The reduction in 25(OH)D_3 may have offset the increase in 25(OH)D_2 levels [28]. However, more research has confirmed that whether in capsule form, added to foods, or obtained from UV-irradiated mushrooms, vitamin D_2 can effectively increase serum levels of 25(OH)D [8,29,30]. Our study shows that after 4 weeks of consuming PC treated with PUV light, the participants had a significant increase in serum 25(OH)D_2 levels, while 25(OH)D_3 levels not only did not decrease but also showed a slight increasing trend. Since we only asked the participants to avoid mushroom-related foods during the experiment and did not place restrictions on foods rich in vitamin D3, such as fish, eggs, and milk, and since we encouraged them to maintain their usual daily activities and food intake, this may have contributed to the maintenance of serum 25(OH)D_3 levels among the participants. Kleftaki et al.'s study demonstrated that a *Pleurotus eryngii* mushroom snack enhanced with vitamin D2 improved glucose regulation, reduced body weight, fat, waist, and hip circumferences, and increased 25(OH)D_2 levels in metabolically unhealthy patients. It also lowered LDL, SGOT, IL-6, and ox-LDL levels while enhancing overall physical health, showcasing antidiabetic, antiobesity, anti-inflammatory, and antioxidant benefits [31].

In this study, we also observed an interesting result: the PC-100 g group exhibited a significant decrease in serum intact PTH (I-PTH) levels. Some studies have observed a negative relationship between 25(OH)D levels and PTH levels [19,32]. According to physiological paradigms, vitamin D promotes calcium absorption to maintain serum calcium levels. When vitamin D is deficient, serum calcium levels may decrease, which stimulates increased PTH secretion to maintain calcium homeostasis, often at the cost of increased bone turnover. However, in this study, all the participants maintained serum calcium levels within the normal range before and after the trial, with no significant changes observed across the groups. Despite the reduction in PTH levels in the PC-100 g group after consuming a high dose of PUV-treated PC, serum calcium levels remained stable. While this aligns with the theoretical relationship between vitamin D and calcium homeostasis, further studies are required to confirm whether PUV-treated PC directly aids calcium absorption or affecpts markers of bone turnover.

Revious animal studies have shown that consuming vitamin-D-enriched edible mushrooms can enhance serum vitamin D levels, lower PTH concentrations, and improve the osteoid area while slowing trabecular bone loss in the femur [33,34]. However, without direct evidence from our study on changes in calcium levels or serum markers of bone formation and resorption, any claims about the impact of PUV-treated PC on calcium homeostasis remain speculative and should be interpreted with caution.

Fulgoni et al. analyzed data from the National Health and Nutrition Examination Survey (NHANES) 2011–2016 and found that incorporating a serving of UV-exposed mushrooms (providing 5 μg of vitamin D per serving) to the daily diet nearly doubles vitamin D intake (an increase of 98–104%). This dietary change significantly reduces the risk of vitamin D deficiency and simultaneously increases the intake of various other nutrients [35]. A dietary model also indicates that consuming four servings per week of UV-exposed mushrooms can help most Australian adults meet their vitamin D recommendations [36]. Therefore, UV-exposed mushrooms represent a crucial tool in the global strategy for addressing vitamin D deficiency.

Initially, our study was designed with 10 g of PC (approximately 20 μg of vitamin D) as the low dose and a maximum of one serving per day (100 g) of mushrooms (approximately 200 μg of vitamin D) as the high dose. The results demonstrated that the high dose more effectively improved vitamin D status, and no adverse side effects were reported among the participants. However, this was a short-term trial, and long-term safety assessments are still needed. Recently, the European Food Safety Authority's Panel on Nutrition, Novel Foods, and Food Allergens (NDA) reviewed the production process, composition, and specifications of *Agaricus bisporus* mushroom powder that had been exposed to ultraviolet (UV) irradiation to induce vitamin D_2 at levels ranging from 245–460 μg/g as a novel food (NF) according to EU regulations. The panel considers that the total intake of vitamin D from novel foods (NFs), background diets, and fortified foods is safe when it is below the upper limits (ULs) for vitamin D previously established by the NDA panel for children, adolescents, and adults (50 μg/day and 100 μg/day, respectively) [37]. This recommendation may also be referenced in the future when promoting PUV-treated PC.

This study has several limitations. The relatively small sample size and short 4-week intervention period restricted the observation of long-term physiological effects. Additionally, challenges from the COVID-19 pandemic, including participant recruitment and follow-up restrictions, further constrained the study's scope and sample size. Another limitation is the approximately 8% loss of vitamin D in the PUV-treated PC mushrooms after steaming, with the preservation of vitamin D using other cooking methods remaining unclear. Furthermore, although reductions in iPTH levels were observed, serum calcium levels in all groups remained within the normal range, indicating stable calcium homeostasis. As a result, this study could not provide conclusive evidence on the effects of PUV-treated PC on calcium metabolism or bone health markers. These limitations highlight the importance of conducting larger-scale and longer-term studies in the future to validate the findings and evaluate the long-term benefits of PUV-treated PC.

Cultivation trials demonstrate that PC mushrooms can thrive under a wide range of conditions, including growth on broad-leaf hardwood sawdust at a pH of 6 and temperatures between 20–32 °C. This adaptability makes them suitable for year-round cultivation across all regions of Taiwan. In the future, PC mushrooms could serve as a natural, sustainable, and fungi-derived food to address vitamin D deficiencies in Taiwan and Asia.

5. Conclusions

Based on the research findings, consuming *Pleurotus citrinopileatus* (PC) treated with PUV effectively increased serum 25(OH)D_2 levels in a dose-dependent manner, improving the participants' vitamin D nutritional status without any observed adverse side effects. These results highlight the potential of PUV-treated PC as a valuable source of vitamin D, contributing to advancements in processing, storage preservation, and addressing vitamin D deficiency-related disease control.

Supplementary Materials: The following supporting information can be downloaded at: https://www.mdpi.com/article/10.3390/foods14020259/s1, Supplementary File S1: CONSORT 2010 checklist of information to include when reporting a randomised trial.

Author Contributions: Conceptualization, S.-C.L. and C.-M.C.; methodology, S.-C.L.; software, Y.-C.W.; validation, C.-T.S., T.-C.F. and W.S.; formal analysis, C.-C.H.; resources, S.-H.Y., Y.-M.J. and H.-M.H.; data curation, C.-C.H.; writing—original draft preparation, C.-C.H. and C.-M.C.; writing—review and editing, C.-M.C. and S.-C.L.; supervision, S.-C.L.; funding acquisition, S.-C.L. All authors have read and agreed to the published version of the manuscript.

Funding: This study was supported by the Council of Agriculture in Taiwan. The funding number is 111AS-11.1.2-CI-C1.

Institutional Review Board Statement: This study was conducted in accordance with the Declaration of Helsinki and approved by the Taipei Medical University Institutional Review Board (approval No. N202111054, approval date: 20 February 2022). Additionally, this study has been registered with ClinicalTrials.gov (registration No. NCT05716698, approval date: 15 April 2022).

Informed Consent Statement: Informed consent was obtained from all subjects involved in the study.

Data Availability Statement: The original contributions presented in this study are included in the article/Supplementary Materials. Further inquiries can be directed to the corresponding author.

Acknowledgments: We sincerely thank all the participants in this study. We also acknowledge the Small Molecule Metabolomics Core Facility, sponsored by IPMB and the Scientific Instrument Center, Academia Sinica, for their technical assistance and support. Additionally, the mass spectrometry analysis was supported by the Academia Sinica Metabolomics Core Facility at the Agricultural

Biotechnology Research Center, under the Academia Sinica Core Facility and Innovative Instrument Project (AS-CFII-111-218).

Conflicts of Interest: The authors declare no conflicts of interest.

References

1. Wimalawansa, S.J. Physiology of Vitamin D—Focusing on Disease Prevention. *Nutrients* **2024**, *16*, 1666. [CrossRef] [PubMed]
2. Christakos, S.; Dhawan, P.; Verstuyf, A.; Verlinden, L.; Carmeliet, G. Vitamin D: Metabolism, molecular mechanism of action, and pleiotropic effects. *Physiol. Rev.* **2016**, *96*, 365–408. [CrossRef] [PubMed]
3. Bikle, D.D. Vitamin D metabolism, mechanism of action, and clinical applications. *Chem. Biol.* **2014**, *21*, 319–329. [CrossRef] [PubMed]
4. Lee, M.J.; Hsu, H.J.; Wu, I.W.; Sun, C.Y.; Ting, M.K.; Lee, C.C. Vitamin D deficiency in northern Taiwan: A community-based cohort study. *BMC Public Health* **2019**, *19*, 337. [CrossRef] [PubMed]
5. Outila, T.A.; Kärkkäinen, M.U.; Seppänen, R.H.; Lamberg-Allardt, C.J. Dietary intake of vitamin D in premenopausal, healthy vegans was insufficient to maintain concentrations of serum 25-hydroxyvitamin D and intact parathyroid hormone within normal ranges during the winter in Finland. *J. Am. Diet. Assoc.* **2000**, *100*, 434–441. [CrossRef]
6. Chan, J.; Jaceldo-Siegl, K.; Fraser, G.E. Serum 25-hydroxyvitamin D status of vegetarians, partial vegetarians, and nonvegetarians: The Adventist Health Study-2. *Am. J. Clin. Nutr.* **2009**, *89*, 1686s–1692s. [CrossRef]
7. Yang, M.Y.; Huang, C.Y.; Chiu, T.H.T.; Chang, K.C.; Lin, M.N.; Chen, L.Y.; Hu, A. Using gas chromatography and mass spectrometry to determine 25-hydroxyvitamin D levels for clinical assessment of vitamin D deficiency. *J. Food Drug Anal.* **2019**, *27*, 494–501. [CrossRef]
8. Keegan, R.J.; Lu, Z.; Bogusz, J.M.; Williams, J.E.; Holick, M.F. Photobiology of vitamin D in mushrooms and its bioavailability in humans. *Dermato-Endocrinology* **2013**, *5*, 165–176. [CrossRef]
9. Phillips, K.M.; Ruggio, D.M.; Horst, R.L.; Minor, B.; Simon, R.R.; Feeney, M.J.; Byrdwell, W.C.; Haytowitz, D.B. Vitamin D and sterol composition of 10 types of mushrooms from retail suppliers in the United States. *J. Agric. Food Chem.* **2011**, *59*, 7841–7853. [CrossRef]
10. Kalaras, M.D.; Beelman, R.B.; Elias, R.J. Effects of postharvest pulsed UV light treatment of white button mushrooms (*Agaricus bisporus*) on vitamin D2 content and quality attributes. *J. Agric. Food Chem.* **2012**, *60*, 220–225. [CrossRef]
11. Cardwell, G.; Bornman, J.F.; James, A.P.; Black, L.J. A review of mushrooms as a potential source of dietary vitamin D. *Nutrients* **2018**, *10*, 1498. [CrossRef] [PubMed]
12. Hao, Y.; Sun, H.; Zhang, X.; Wu, L.; Zhu, Z. A novel polysaccharide from *Pleurotus citrinopileatus* mycelia: Structural characterization, hypoglycemic activity and mechanism. *Food Biosci.* **2020**, *37*, 100735. [CrossRef]
13. Raman, J.; Jang, K.-Y.; Oh, Y.-L.; Oh, M.; Im, J.-H.; Lakshmanan, H.; Sabaratnam, V. Cultivation and nutritional value of prominent *Pleurotus* spp.: An overview. *Mycobiology* **2021**, *49*, 1–14. [CrossRef] [PubMed]
14. Huang, S.-J.; Lin, C.-P.; Tsai, S.-Y. Vitamin D2 content and antioxidant properties of fruit body and mycelia of edible mushrooms by UV-B irradiation. *J. Food Compos. Anal.* **2015**, *42*, 38–45. [CrossRef]
15. Tidke, S.; Pooja, B.; Manjula, B.; Singh, R.; Shora, R.; Arush, H.; Sinosh, S.; Geethanath, S.; Kiran, S.; Ravishankar, G. Enhancement of nutraceutical components of mushroom by uv exposure and extension of their shelf-life using edible coating material adopting online assessment of keeping quality by Magnetic Resonance Imaging (MRI) technique. *Appl. Chem. Eng.* **2024**, *7*, 1–10. [CrossRef]
16. Rathore, H.; Prasad, S.; Sehwag, S.; Sharma, S. Vitamin D 2 fortification of *Calocybe indica* mushroom by natural and artificial UVB radiations and their potential effects on nutraceutical properties. *3 Biotech* **2020**, *10*, 41. [CrossRef]
17. Judprasong, K.; Chheng, S.; Chimkerd, C.; Jittinandana, S.; Tangsuphoom, N.; Sridonpai, P. Effect of Ultraviolet Irradiation on Vitamin D in Commonly Consumed Mushrooms in Thailand. *Foods* **2023**, *12*, 3632. [CrossRef]
18. Koyyalamudi, S.R.; Jeong, S.-C.; Pang, G.; Teal, A.; Biggs, T. Concentration of vitamin D2 in white button mushrooms (*Agaricus bisporus*) exposed to pulsed UV light. *J. Food Compos. Anal.* **2011**, *24*, 976–979. [CrossRef]
19. Chen, C.M.; Mu, S.C.; Chen, Y.L.; Tsai, L.Y.; Kuo, Y.T.; Cheong, I.M.; Chang, M.L.; Li, S.C. Infants' Vitamin D Nutritional Status in the First Year of Life in Northern Taiwan. *Nutrients* **2020**, *12*, 404. [CrossRef]
20. Giustina, A.; Lazaretti-Castro, M.; Martineau, A.R.; Mason, R.S.; Rosen, C.J.; Schoenmakers, I. A view on vitamin D: A pleiotropic factor? *Nat. Rev. Endocrinol.* **2024**, *20*, 202–208. [CrossRef]
21. Cui, A.; Zhang, T.; Xiao, P.; Fan, Z.; Wang, H.; Zhuang, Y. Global and regional prevalence of vitamin D deficiency in population-based studies from 2000 to 2022: A pooled analysis of 7.9 million participants. *Front. Nutr.* **2023**, *10*, 1070808. [CrossRef] [PubMed]
22. Kim, K.N.; Lee, J.S.; Shim, J.S.; Yoon, M.O.; Lee, H.S. Estimated dietary vitamin D intake and major vitamin D food sources of Koreans: Based on the Korea National Health and Nutrition Examination Survey 2016–2019. *Nutr. Res. Pract.* **2023**, *17*, 451–463. [CrossRef] [PubMed]

23. Lee, M.S.; Li, H.L.; Hung, T.H.; Chang, H.Y.; Yang, F.L.; Wahlqvist, M.L. Vitamin D intake and its food sources in Taiwanese. *Asia Pac. J. Clin. Nutr.* **2008**, *17*, 397–407. [PubMed]
24. Jiang, W.; Wu, D.-B.; Xiao, G.-B.; Ding, B.; Chen, E.-Q. An epidemiology survey of vitamin D deficiency and its influencing factors. *Med. Clin.* **2020**, *154*, 7–12. [CrossRef] [PubMed]
25. Aditya; Neeraj; Bhatia, J. Escalating Vitamin D2 Content in Fresh Sporocarps of Elm Oyster Mushroom *Hypsizygus ulmarius* (Bull.) Redhead (Agaricomycetes) Using Ultraviolet Irradiation and Assessment of Its Stability, Antioxidant Capacity, Color, and Textural Properties. *ACS Food Sci. Technol.* **2024**, *4*, 2739–2751. [CrossRef]
26. Kalaras, M.D.; Beelman, R.B.; Holick, M.F.; Elias, R.J. Generation of potentially bioactive ergosterol-derived products following pulsed ultraviolet light exposure of mushrooms (*Agaricus bisporus*). *Food Chem.* **2012**, *135*, 396–401. [CrossRef]
27. Urbain, P.; Singler, F.; Ihorst, G.; Biesalski, H.K.; Bertz, H. Bioavailability of vitamin D_2 from UV-B-irradiated button mushrooms in healthy adults deficient in serum 25-hydroxyvitamin D: A randomized controlled trial. *Eur. J. Clin. Nutr.* **2011**, *65*, 965–971. [CrossRef]
28. Stephensen, C.B.; Zerofsky, M.; Burnett, D.J.; Lin, Y.P.; Hammock, B.D.; Hall, L.M.; McHugh, T. Ergocalciferol from mushrooms or supplements consumed with a standard meal increases 25-hydroxyergocalciferol but decreases 25-hydroxycholecalciferol in the serum of healthy adults. *J. Nutr.* **2012**, *142*, 1246–1252. [CrossRef]
29. Biancuzzo, R.M.; Young, A.; Bibuld, D.; Cai, M.H.; Winter, M.R.; Klein, E.K.; Ameri, A.; Reitz, R.; Salameh, W.; Chen, T.C.; et al. Fortification of orange juice with vitamin D(2) or vitamin D(3) is as effective as an oral supplement in maintaining vitamin D status in adults. *Am. J. Clin. Nutr.* **2010**, *91*, 1621–1626. [CrossRef]
30. Holick, M.F.; Biancuzzo, R.M.; Chen, T.C.; Klein, E.K.; Young, A.; Bibuld, D.; Reitz, R.; Salameh, W.; Ameri, A.; Tannenbaum, A.D. Vitamin D2 is as effective as vitamin D3 in maintaining circulating concentrations of 25-hydroxyvitamin D. *J. Clin. Endocrinol. Metab.* **2008**, *93*, 677–681. [CrossRef]
31. Kleftaki, S.-A.; Amerikanou, C.; Gioxari, A.; Lantzouraki, D.Z.; Sotiroudis, G.; Tsiantas, K.; Tsiaka, T.; Tagkouli, D.; Tzavara, C.; Lachouvaris, L.; et al. A Randomized Controlled Trial on *Pleurotus eryngii* Mushrooms with Antioxidant Compounds and Vitamin D2 in Managing Metabolic Disorders. *Antioxidants* **2022**, *11*, 2113. [CrossRef]
32. Malik, M.Z.; Latiwesh, O.B.; Nouh, F.; Hussain, A.; Kumar, S.; Kaler, J. Response of Parathyroid Hormone to Vitamin D Deficiency in Otherwise Healthy Individuals. *Cureus* **2020**, *12*, e9764. [CrossRef]
33. Malik, M.A.; Jan, Y.; Al-Keridis, L.A.; Haq, A.; Ahmad, J.; Adnan, M.; Alshammari, N.; Ashraf, S.A.; Panda, B.P. Effect of Vitamin-D-Enriched Edible Mushrooms on Vitamin D Status, Bone Health and Expression of CYP2R1, CYP27B1 and VDR Gene in Wistar Rats. *J. Fungi* **2022**, *8*, 864. [CrossRef]
34. Won, D.J.; Seong, K.S.; Jang, C.H.; Lee, J.S.; Ko, J.A.; Bae, H.; Park, H.J. Effects of vitamin D(2)-fortified shiitake mushroom on bioavailability and bone structure. *Biosci. Biotechnol. Biochem.* **2019**, *83*, 942–951. [CrossRef]
35. Fulgoni, V.L., 3rd; Agarwal, S. Nutritional impact of adding a serving of mushrooms on usual intakes and nutrient adequacy using National Health and Nutrition Examination Survey 2011–2016 data. *Food Sci. Nutr.* **2021**, *9*, 1504–1511. [CrossRef]
36. Starck, C.; Cassettari, T.; Wright, J.; Petocz, P.; Beckett, E.; Fayet-Moore, F. Mushrooms: A food-based solution to vitamin D deficiency to include in dietary guidelines. *Front. Nutr.* **2024**, *11*, 1384273. [CrossRef]
37. Turck, D.; Bohn, T.; Castenmiller, J.; De Henauw, S.; Hirsch-Ernst, K.I.; Maciuk, A.; Mangelsdorf, I.; McArdle, H.J.; Naska, A.; Pentieva, K.; et al. Safety of vitamin D(2) mushroom powder as a Novel food pursuant to Regulation (EU) 2015/2283 (NF 2020/2226). *EFSA J.* **2024**, *22*, e8817. [CrossRef]

foods

MDPI

Review

Mushrooms as Nutritional Powerhouses: A Review of Their Bioactive Compounds, Health Benefits, and Value-Added Products

Akruti Singh [1], Ramesh Kumar Saini [1], Amit Kumar [1], Prince Chawla [2] and Ravinder Kaushik [1,*]

[1] School of Health Sciences and Technology, UPES, Dehradun 248007, Uttarakhand, India; akruti.singh0898@gmail.com (A.S.); rameshkumar.saini@ddn.upes.ac.in (R.K.S.); kumar.amit@ddn.upes.ac.in (A.K.)

[2] Department of Food Technology and Nutrition, Lovely Professional University, Phagwara 144411, Punjab, India; princefoodtech@gmail.com

* Correspondence: ravinder_foodtech2007@rediffmail.com

Abstract: Mushrooms are known to be a nutritional powerhouse, offering diverse bioactive compounds that promote and enhance health. Mushrooms provide a distinguishable taste and aroma and are an essential source of vitamin D_2, vitamin B complex, hydroxybenzoic acids (HBAs) and hydroxycinnamic acids (HCAs), terpenes, sterols, and β-glucans. Edible mushroom varieties such as *Hericium erinaceus*, *Ganoderma* sp., and *Lentinula edodes* are recognized as functional foods due to their remarkable potential for disease prevention and promotion of overall health and well-being. These varieties have antioxidants, anti-inflammatory, cytoprotective, cholesterol-lowering, antidiabetic, antimicrobial, and anticancer properties, as well as controlling blood pressure, being an immunity booster, and strengthening bone properties. In addition, they contain essential non-digestible oligosaccharides (NDOs) and ergothioneine, a potential substrate for gut microflora. Supplementing our daily meals with those can add value to our food, providing health benefits. Novel edible mushrooms are being investigated to explore their bioactive substances and their therapeutic properties, to benefit human health. The scientific community (mycologists) is currently studying the prospects for unlocking the full health advantages of mushrooms. This review aims to promote knowledge of mushroom culturing conditions, their nutritional potential, and the value-added products of 11 varieties.

Keywords: functional foods; edible fungi; bioactive compounds; health benefits; value-added products

1. Introduction

The Indian system of medicine, Ayurveda, emphasizes that "When diet is wrong, medicine is of no use. When diet is correct, medicine is of no need". In this context, mushrooms can be vital in providing a balanced and wholesome diet [1]. Mushrooms, a kind of edible fungi that form sizable, firm, or robust fleshy structures, are found abundantly across the globe. Approximately 16,000 different types of edible mushrooms have been identified. Nearly 7000 varieties are recognized for their excellent taste and nutritional profile, and nearly 3000 are regularly included in daily food menus [2]. Basidiomycetes, particularly those found in the order Agaricales, are among the most notable in the diverse world of mushrooms. The structure of mushrooms includes several key components: mycelium, hypha, cap, lamellae, spores, stem, voula, and rings [3]. Mushrooms have an excellent nutritional profile and so, for centuries, mushrooms have been integrated into meals and their medicinal benefits for overall well-being leveraged [4]. It is estimated that there are nearly 1.5 million different species of fungi, among which scientists have identified around 110,000 types. Numerous mushrooms offer a wealth of flavor and essential nutrients; however, caution is paramount as some species are poisonous [5]. Mushrooms have historically been used in Traditional Chinese Medicine, for medicinal purposes for 3000 and 7000 years. For example, shiitake mushrooms, scientifically known as *L. edodes* (Berk.) Pegler, have been utilized both in nutrition and medicine since 600–1000 B.C. [6]. These fungi produce bioactive compounds such as peptides, sterols, polysaccharides, proteins, and phenols, which can be considered potential drugs [7]. Due to their rich nutritional profile and organoleptic properties, mushrooms are often blended into various dishes and can also serve as a meat substitute [8]. The nutritional profile is significantly influenced by the substrate a mushroom feeds on, environmental parameters, and its stage of maturity. Mushrooms contain various carbohydrates such as glycogen, xylose, mannose, galactose, glucose, and some insoluble forms like fiber, mannan, cellulose, and chitin. They also have a valuable compound known as glucan, which is characterized by glycosidic bonds at β (1, 3), β (1, 4), and β (1, 6), making mushrooms an excellent addition to a healthy diet [9]. With a unique umami flavor, mushrooms are consumed as part of everyday food dishes while enhancing their nutritional value [10]. As Kaul states, "Medicines and food have a common origin" [11].

This review aims to promote complete knowledge of mushroom culturing conditions and the nutritional potential of different varieties like *Agaricus bisporus, Calocybe indica, Volvariella volvacea, Auricularia polytricha, Schizophyllum commune, G. lucidium, Pleurotus ostreatus, Grifola frondosa, H. erinaceus, Flammulina velutipes,* and *L. edodes*. Their health benefits include anti-inflammatory, antioxidant, antidiabetic, antimicrobial, enhanced gut microbiota, and healing properties. In addition, we explore the mushroom value-added products available on the market. By consolidating information about the structures, bioactive compounds, and diverse uses of mushrooms, this article underscores the importance of mushrooms as a unique and valuable food source, which contributes to overall health and well-being.

2. Culturing Conditions

Mushroom fructification, the process of producing fruiting bodies, is initiated by a mature mycelia network. Many mushrooms are saprophytic fungi, acting as crucial decomposers in diverse ecosystems. For successful cultivation, the mushroom-growing conditions must be carefully optimized, and the substrate composition and formulation vary according to the species being cultivated [12]. Researchers are increasingly using agro-industrial wastes as substrates, including agricultural and industrial residues with low nitrogen contents [13,14]. Substrates also include organic materials such as cereal by-products (bran and shell), soybean meal, and compost, as well as inorganic supplements like ammonium salt and fertilizers, which provide the necessary nitrogen for mushroom cultivation [15,16]. Additionally, pulses, maize, soybean, sorghum, and residues from oil seeds, sugarcane, and cotton can be utilized as organic substrates. Even sawed wood residues are well-known substrates supporting mushroom growth and fruiting [10]. Some studies have utilized tea waste as a substrate for cultivating oyster mushroom varieties [17]. The roles of intrinsic and extrinsic factors are of equal significance in mushroom cultivation. Intrinsic factors include the carbon and nitrogen contents of the growth medium, pH level, and appropriate nutrition media, while extrinsic factors encompass temperature, humidity, light, and gas concentrations (CO_2) [18–20]. The cultivation process enhances the economic use of agricultural residue to produce mushrooms and improves the relationships between fungal hyphae, substrates, and soil systems [21]. Several studies explored the use of synthetic or semisynthetic media as substrates for mushroom mycelium cultivation. Artificially synthesized media can supply the necessary nutrients for growth, including SDA (Sabouraud dextrose agar), MYE (malt yeast extract), YMEA (yeast malt extract agar), and PDA (potato dextrose agar). Additionally, enriched potato culture media like YPDA (yeast potato dextrose agar), PMA (potato malt agar), CMA (corn meal agar), PDYA (potato dextrose yeast agar), PM (potato malt peptone), PCA (potato carrot agar), PGA (potato glucose agar), and PSG (potato sucrose gelatin) have been noted in studies as effective substrates for mycelium culturing. The analysis showed that synthetic media such as PDA and MEA provide the maximum nutrient effect due to their rich nutrient composition, which is essential for optimal fungal growth. The influence of culture media on mycelia development varies based on the fungal species and strains. Nutrient media like PDA or PGA, followed by MEA and MCM, were identified as the best options for promoting mycelia development. However, agriculture and food wastes contain a significant number of natural-based chemicals, which can result in greater biomass growth than synthetic and semi-synthetic media. The optimal cultivation temperature for mycelia is often linked to the fungus's genetic origin and the environmental conditions in which it naturally grows. Most basidiomycetes thrive at temperatures between 20 and 30 °C, while some species prefer higher temperatures ranging from 35 to 37 °C [14,19,22].

3. Nutritional Potential of Mushroom

Edible mushrooms are grown with a mini packet of essential nutrients, which include a good amount of water, carbohydrates, protein, lipids, fibers, macronutrients, and micronutrients [23,24]. Mushrooms contain both primary and secondary metabolites. Primary metabolites are responsible for energy production, while secondary metabolites are responsible for medicinal properties [1]. These macrofungi's organoleptic features and high nutritional content contribute to their growing popularity. Edible mushrooms are rich in proteins and account for 19–35% of the dry mass, while carbohydrates constitute 50–65% of the dry mass, rendering mushrooms an abundant source of high-quality dietary fiber. Furthermore, mushrooms exhibit a low lipid content, ranging from 2% to 6% of the dry mass, and are regarded as hypocaloric [25]. These powerful compounds provide numerous health benefits including antimicrobial defense, protection against oxidative damage, and anti-inflammation properties. Moreover, mushrooms exhibit antidiabetic, anticancer, antiviral, and anti-immunomodulatory activities, making them valuable ingredients in the development of functional foods [26–28]. Mushrooms also contain many bioactive compounds including alkaloids, ergosterols, polysaccharides, polyphenols, terpenoids, lectins, glycoproteins, sesquiterpenes, sterols, and lactones. The concentrations of these bioactive chemicals vary considerably based on several parameters, including culture, strain, storage conditions, substrate, and processing conditions [29]. The nutritional compositions of some edible mushrooms are presented in Table 1.

Table 1. Nutritional profiles of different mushroom varieties.

Mushroom Varieties	Content/100 g					Reference
	Carbohydrate	Protein	Fat	Fiber	Ash	
Agaricus bisporus	47.20	32.10	3.10	8.90	8.70	[30]
Auricularia polytricha	51.65	18.7	1.60	22.80	3.77	[31,32]
Volvariella volvacea	43.16	19.40	2.49	15.10	11.71	[33]
Calocybe indica	6.80	3.22	1.05	1.11	2.30	[34]
Schizophyllum commune	42.0	15.55	9.00	30.00	3.50	[35]
Ganoderma lucidium	37.33	8.54	1.91	50.19	2.03	[36]
Pleurotus ostreatus	61.10	20.00	2.50	7.90	7.70	[30]
Flammulina velutipes	42.83	20.26	4.50	23.31	9.00	[37]

Table 1. *Cont.*

Mushroom Varieties	Content/100 g					Reference
	Carbohydrate	Protein	Fat	Fiber	Ash	
Grifola frondosa	70.70	19.30	3.80	-	6.10	[38]
Lentinula edodes	64.40	22.80	2.10	1.26–2.95	6.00	[39]
Hericium erinaceus	57.00	22.30	3.50	3.30–7.80	-	[40]

3.1. Carbohydrates

Mushrooms are low-calorie foods owing to low carbohydrate contents, minimal sugar levels (no glucose), and high fiber contents. They contain various carbohydrates including simple sugars like sucrose, xylose, glycogen, rhamnose, mannose, fructose, galactose, mannose, and xylose, and polysaccharides such as cellulose, glycoproteins, α-glucans, and β-glucans, glucan, mannoglucan, heteroglycan, galactomannan, and lentinan [41,42]. Mushrooms are also high in dietary fibers, primarily non-starch polysaccharides, with 4 to 9% being soluble and 22 to 30% insoluble [43]. They contain non-digestible carbohydrates, such as chitin and (1→3)-β-d-glucans, which promote intestinal health, and the main components of the cell wall are β-glucans and polysaccharides. Additionally, mushrooms contain non-digestible oligosaccharides (NDOs) consisting of carbohydrate molecules of fewer than 20 monosaccharide units joined by glycosidic linkages. These NDOs are resistant to hydrolysis by salivary and intestinal digestion enzymes associated with various beneficial advantages, including antipathogenic and prebiotic characteristics. Individuals can increase their intake of NDO through food sources, like mushrooms, and supplements derived from dried fruiting bodies or mycelium-based products from fungal species [44–46]. Although mushrooms have less fiber than vegetables and fruits, they are still a nutritious, low-energy dietary option, particularly beneficial for type II diabetes and those seeking weight loss. Due to their low glycemic index (GI) and glycemic load (GL), they do not cause spikes in blood sugar levels [47,48]. In *L. edodes*, key polysaccharides such as emitanin, lentinan (a β-(1,3)-D-glucan enhances the effectiveness of chemotherapeutic drugs), and KS-2 are found to benefit health [49].

3.2. Protein

Edible mushrooms are often high in protein, although the protein content varies widely depending on the mushroom's species, stage of growth, and growth medium. They contain essential amino acids such as lysine, valine, tryptophan, isoleucine, methionine, leucine, and threonine. Mushrooms also provide proteins such as lectins, laccases, histidine, phenylalanine, and cysteine [50–53]. Notable mushrooms of the *Pleurotus* species have high-quality protein due to the effective distribution of essential and non-essential amino acids such as gamma-aminobutyric acid (GABA), a critical neurotransmitter [50]. When compared to other food sources, the protein content of edible mushrooms is quite competitive [54]. Animal-based foods (dry weight) contain protein, which is present in proportions of at least 27% for milk, 37–83% for meat, 53% for eggs, and the highest 58–90% for fish and crustaceans [55]. In contrast, plant-based sources such as legumes contain 22–40%, cereals 8–18%, nuts 4–20%, other seeds 18–32%, and tubers less than 10% [56,57]. Certain species of edible fungi, *A. bisporus* 32.10%, *H. erinaceus* 22.30%, and *L. edodes* 22.80%, provide protein concentrations that are equal to or exceed those found in animal-derived sources such as dairy products, meat, eggs, and seafood [54,58–61]. Consequently, these edible fungi represent an exceptional source of high-quality protein that is more accessible, cost-effective, and exhibits a reduced environmental footprint, and so, in the future, they will become a compelling alternative to both animal-derived and various plant-based protein options [62]. Ergothioneine (EGT) is uniquely sulfur-containing and has excellent free radical scavenging activity. EGT is found in high concentrations in mushroom species like hen of the woods, shiitake, King Boletes, Enokitake, and oyster mushrooms [62–65]. It has been correlated with several health benefits, including lower rates of dementia and cardiovascular disease and anti-inflammatory and cytoprotective effects, and it may even lead to a longer life expectancy. Mushrooms have much greater quantities of ergothioneine than cereals, vegetables, and meat [64,66]. As such, edible mushrooms are appealing foods with significant nutritional benefits that contribute to overall health.

3.3. Fats

Edible mushrooms represent a low-calorie aliment with a minimal fat content (4–6%). *A. bisporus*, also known as the button mushroom, has a total fat content ranging from 0.34 to 2.2 g per 100 g of dry weight [67,68]. The three major fatty acids present in edible mushrooms are linoleic acid (C18:2), oleic acid (C18:1), and palmitic acid (C16:0). Linoleic acid is useful in reducing the amount of lipids in the blood and helping to alleviate arthritis symptoms [50]. Additionally, these fungi are abundant in polyunsaturated fatty acids (PUFAs), particularly oleic (1.1–12.3 g/100 g fresh weight (FW)), stearic (1.6–3.1 g/100 g FW), palmitic (10.3–11.9 g/100 g FW), and linoleic acids. Ergosterol (ergosta-5,7,22-trien-3b-ol) is identified as the most prevalent sterol within edible mushrooms [68]. Although mushrooms are characterized by low caloric and fat contents, they exhibit a markedly elevated ratio of polyunsaturated fatty acids in comparison to saturated fatty acids [69].

3.4. Micronutrients

Mushrooms are rich in various vitamins such as vitamin B complex (B_1, B_2, B_3, B_9, and B_{12}), vitamin C, vitamin D_2, and vitamin E and minerals like calcium, cadmium, magnesium, phosphorus, iron, sodium, cobalt, zinc, potassium, copper, titanium, selenium, and molybdenum [50,70,71]. Tocopherol (α, β, γ, and δ) is a vitamin present in various mushroom varieties [24,71,72]. Mushrooms are known for their high potassium content and low sodium content, as potassium reduces tension in blood vessels and eventually helps in lowering blood pressure [70]. *A. bisporus* is particularly high in Na, Li, and K, but poor in Cu, Mn, Cr, Co, Pb, Ni, and Zn [73]. *H. erinaceus* contain high levels of K, P, and Mg followed by Na, Fe, Ca, Zn, Al, Cu, Li, Mn, and Ba. *G. lucidium* contains high levels of K, P, and Ca, followed by Mg, Na, Fe, Al, B, Zn, and Cu, and the least Mn [74].

3.5. Bioactive Compounds

Mushrooms encompass a diverse array of bioactive constituents, which include phenolic acids, glycosides, volatile substances, alkaloids, flavonoids, organic acids, and a variety of biological catalysts such as amylases, cellulases, laccases, lipases, pectinases, proteases, phytases, and xylanases. The phenolic constituents identified within mushrooms comprise gallic acid, p-coumaric acid, caffeic acid, p-hydroxybenzoic acid, protocatechuic acid, and pyrogallol [75,76]. The majority of phenolic acids in mushrooms are hydroxybenzoic acids (HBAs) and hydroxycinnamic acids (HCAs). HBAs may be found in complex compounds such as tannins, lignin, and organic acids, whereas HCAs are attached to cell wall components such as lignin, cellulose, and protein. The most prevalent HCAs encountered in mushrooms include ferulic, sinapic, caffeic, and p-/o-coumaric acids, which play critical roles in lignin biosynthesis, disease resistance, and growth regulation [77,78]. Mild alkaline hydrolysis is the most effective for extracting them. In mushrooms, mainly quinic acid esters, and also gallic, gentisic, homogentisic, p-hydroxybenzoic, protocatechuic, 5-sulphosalicylic, syringic, vanillic, and veratric, are the most often observed HBA derivatives [24,71,72]. Anthocyanidins, biochanin, flavanols, flavones, isoflavones, flavanones, catechin, chrysin, myricetin, hesperetin, naringenin, naringin, formometin, resveratrol, quercetin, pyrogallol, rutin, and kaempferol are among the flavonoids present in mushrooms [72]. The structures of active compounds in mushrooms are presented in Figure 1. The quantity of these bioactive chemical compounds in mushrooms depends on the substrate, culturing conditions, storage conditions, and cooking procedures [48]. *G. lucidum* produces several terpene derivatives, including ganoderal, ganoderic acids, lucidone, ganodermanondiol, ganodermic, and ganodermanontriol [49]. *H. erinaceus* is known for hericenones (A-J, I, L, and K), erinacines (A-K, P-V, Z1, and Z2) [79,80], and *L. edodes* contains a polysaccharide known as lentinan [81]. The bioactive molecules, health benefits, and food products prepared from mushroom varieties are presented in Table 2.

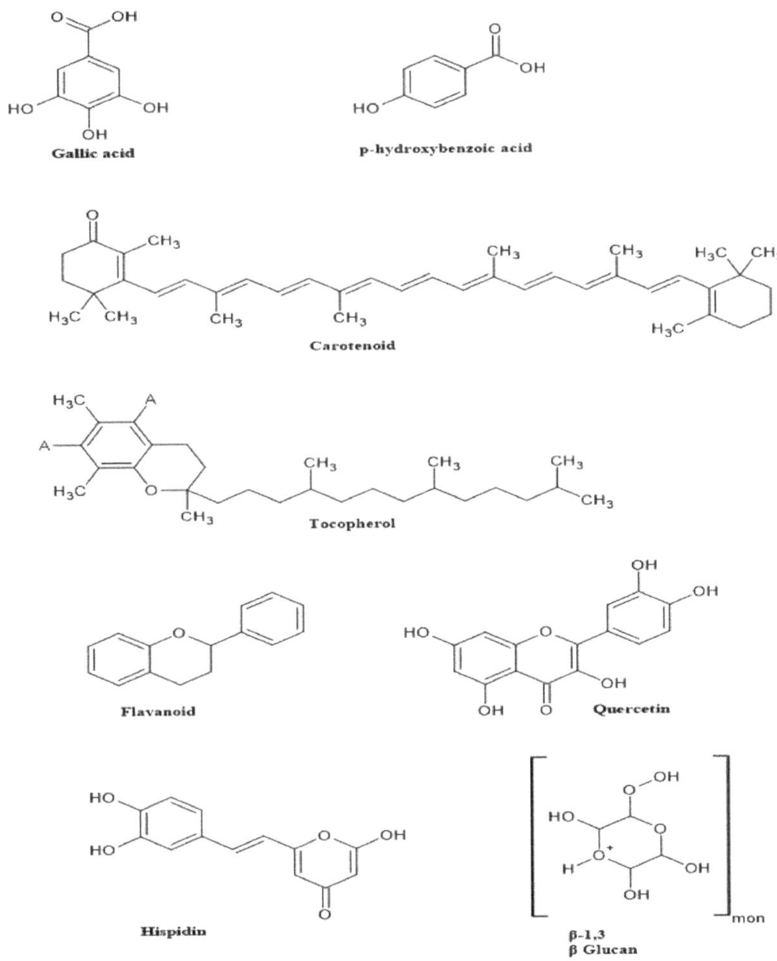

Figure 1. Structures of active compounds in the mushroom [71,81].

Table 2. Bioactive molecules, health benefits, and food products prepared from mushroom varieties.

Mushroom Name	Common Name	Compounds	Health Benefits	Products	References
Agaricus bisporus	Button mushroom	Dietary fiber (chitin), sterols, amino acids, unsaturated fatty acids (linoleic and linolenic acids), phenolic acid (Ferulic acid, gallic acid, cinnamic acid, myricetin, caffeic acid, catechins, protocatechuic acid, p-coumaric acid, chlorogenic, sinapic, *p*-hydroxybenzoic, vanillic, salicylic, and syringic acids), and vitamins (vitamin B complex—vitamin B_1, vitamin B_2, vitamin B_3, vitamin B_6, vitamin B_7, and vitamin B_{12}; vitamin C; vitamin D2; vitamin E (γ-tocopherol, α-tocopherol, δ-tocopherol)), minerals (copper (Cu), cobalt (Co), iron (Fe), selenium (Se), potassium (K), manganese (Mn), phosphorus (P), calcium (Ca), zinc (Zn), iron (Fe), magnesium (Mg), and sodium (Na)), and indols (indole acetic acid, tryptamine, kynurenic acid, melatonin, serotonin and L–tryptophan)	Immunomodulatory and hepatoprotective activities are antimicrobial, antioxidant, antibacterial, antidiabetic, antitumor, anti-inflammatory, antihypertensive, anticancer, and antihypercholesterolemic	Medicinal, cosmetic purposes (lotions, creams, shampoos), functional bread, biscuits	[68,82–86]
Auricularia polytricha	Wood ear/Jew's ear	Phytochemicals, glucan, rhamnose, fucose, arabinose, xylose, mannose, galactose, glucose, and antioxidants	Antioxidant, antitumor, antimicrobial, anti-inflammatory, and immunity boosting	Noodles, steamed white bread	[87–89]
Volvariella volvacea	Straw mushroom/paddy straw mushroom/Chinese mushroom	Beta-glucan, amino acids (leucine, isoleucine, lysine, methionine, tryptophane, valine, threonine, histidine, and phenylalanine), saturated fatty acids and unsaturated fatty acids, vitamins (vitamin C, vitamin B_1, vitamin B_2, vitamin B_3, and vitamin B_7), phenolic acids, tannins and flavonoids, terpenes, polypeptides, and steroids	Anti-inflammatory, hyperlipidemia treatment, anticancer, antioxidant, prevention of chronic hepatitis, antimalarial, antiallergic, and treatment of arteriosclerosis, cardiovascular diseases, and neurodegenerative diseases	Brown rice extruded snacks	[69,90]
Calocybe indica	Paddy straw milky mushroom	Includes 15 amino acids (most abundant glutamic acid and isoleucine), polyphenols (flavonoids, alkaloids, and triterpenoids), phenolics (gallic, vanillin, protocatechuic acid, naringin, naringenin, homogentisic acid, hesperetin, ferulic acid, caffeic acid, and formononetin), minerals (phosphorous (P), magnesium (Mg), barium (Ba), iron (Fe), aluminum (Al), manganese (Mn), copper (Cu), zinc (Zn), boron (Bo), nickel (Ni), chromium (Cr)), 17 fatty acids, vitamins (vitamin B, vitamin D, vitamin C, vitamin E), and β-glucans	Antiaging, physical injury treatment, infection control, cardiovascular disease treatment, antioxidant activity, antilipid peroxidation characteristics, antimicrobial, anticancer, antidiabetic, and immunomodulatory	-	[40,71,91,92]

Table 2. *Cont.*

Mushroom Name	Common Name	Compounds	Health Benefits	Products	References
Schizophyllum commune	Split gill pushroom	Polysaccharides (uronic acid) and monosaccharides (glucose, fucose, ribose, rhamnose, galacturonic acid, galactose, xylose, arabinose, and glucuronic acid)	Anti-inflammatory, immunomodulator, anticancer, and antioxidant	-	[93,94]
Ganoderma lucidium	Lingzhi or Reishi	Organic germanium, heteroglycans, amino acids, glycans, proteoglycans, steroids, polysaccharides, triterpenoids, ganodermanondiol, nucleotides, minerals, vitamins, adenosine beta-d-glucan, ganoderic acids, polyphenols, triterpenoids ergosterol, peroxide ganopoly, ganoderol, hetero galactan-protein complex, lucidimins A-D, lucidimine E, MD-fraction Protein of LZ-8, and phenols (kaempferol, hesperetin, trans-cinnamic acid, gallic acid, quercetin, and naringenin)	Antioxidant, anti-inflammatory, antiallergic, immunomodulating, antitumor properties, diabetes treatment, hepatitis treatment, hypoglycemic, sedative, antiviral (HIV-1), antidepressant, antihepatotoxic, antiosteoporosis, cholesterol biosynthesis inhibitor, dizziness and insomnia treatment, antiobesity, hepatoprotective activity, antidyslipidemia, cardioprotective, neuroprotective, antiepileptic, nootropic, anxiolytic, and radioprotective	Enriched yoghurt, fat replacer to improve cakes, instant Reishi herbal mushroom tea, moon juice spirit dust, *Ganoderma* cell-repairing antiaging face mask, night cream	[11,40,71,72,94–99]
Pleurotus ostreatus	Oyster mushroom/Dhingri/wood fungus	Functional compound beta-glucan, nucleosides, proteins, lectins, polysaccharides, lipopolysaccharides peptides, glycoproteins, triterpenoids, lipids, phenolics (homogentisic, p-Coumaric, cinnamic acids), vitamin B, and vitamin C	Lowers cholesterol, cardiovascular disease treatment, boosts immunity, antitumor, anticarcinogenic, antimicrobial, antioxidant, hepatoprotective, antibacterial, antidiabetic, antiarthritic, and antiviral	Capsule, breadsticks, noodles	[68,71,82,83,100]
Flammulina velutipes	Winter mushroom/Enokitake/golden needle mushroom/velvet stem/lily mushroom/shank mushroom among different countries	Phenolic, sterols, protein, flavonoids, sesquiterpenoids (terpenoids), polysaccharides, glucan complex, lectins, peroxidases, proteases, and isoflavones	Anti-inflammatory, antibacterial, lowering cholesterol, immunomodulating, antiviral, antitumor, antioxidant, antihypertensive, and antifungal	Capsules, meat products (goat meat nuggets), mutton nuggets	[82,83,101,102]
Grifola frondosa	Maitake in Japanese/hen of the woods/hui-shu-hua in Chinese	Polysaccharides, beta-glucans, linoleic acid, triacylglycerols, ergosterol, steroids, homoglucans, heteropolysaccharide, D fraction ergothioneine, grifolan, lectins, oleic acid, and cyanhydric acid	Antitumor, antivirus, hepatoprotection, reducing blood lipid activities, immunomodulation, antioxidation, antiaging, and antihyperglycemia	-	[82,83,86,103]

Table 2. *Cont.*

Mushroom Name	Common Name	Compounds	Health Benefits	Products	References
Lentinula edodes	Shiitake	Beta-D-glucan, polysaccharide, lentin, adenine derivates agaritine, ergothioneine, eritadenine formaldehyde, galactose glucose, lectins, lentinamyicin, lentinan, mannose, polyisoprenoid alcohols RNA from spores statins, phenolics (p-hydroxybenzoic, gallic, 3,4 dihydroxybenzoicacids), vitamins (vitamin B_2, vitamin B_9, vitamin B_1, vitamin B_3, and vitamin B_5), and minerals (silver (Ag), aluminum (Al), cadmium (Cd), chromium (Cr), cesium (Cs), copper (Cu), iron (Fe), mercury (Hg), potassium (K), magnesium ((Mg), molybdenum (Mo), lead (Pb), selenium (Se), zinc (Zn), and calcium (Ca))	Antibiotic, antiviral, potential agent in metabolic syndrome, hyperlipidemia treatment, depressed immune function (including AIDS), anticarcinogenic, anticancer, antiallergic, antifungal, antiviral, bronchial inflammation, heart disease treatment, controlling hypertension, antidiabetes, hepatitis treatment, and regulating urinary inconsistencies	Capsules, muffins	[69,71,82,86,100,103]
Hericium erinaceus	Lion's mane mushroom/Yamabushitake	Glycans, glycoproteins, ergosterol, polysaccharides, hericenones, erinacines, proteins and peptides, beta-glucan, hericenones I, J and 3-hydroxyhericenone F, hericenone L and K, erinacerins A, B, M, and N, erinarol, cyathane-type diterpenoids, octadecenoic acid, vitamins (carotenoids, vitamin C, vitamin E, and vitamin D), lectins, monochlorobenzenes compounds, 4-pyranones, and statins	Antioxidative, anti-inflammatory, anticancer, antimicrobial, antihyperglycemic properties, treatment of neurodegenerative diseases (Parkinson's disease), blood lipid-lowering, immunostimulant, hypoglycemic activity, antidiabetic, and antifatigue	Noodles, fermented juice	[40,58,60,85,103–106]

4. Therapeutic Efficacy of Mushrooms

Medicinal mushrooms are rich in bioactive compounds such as phenolic acids, lectins, β-glucans, polysaccharides, and terpenoids, offering several health benefits that can significantly enhance the quality of life [107,108]. These compounds possess a wide range of properties including prebiotic, immune-modulating, antioxidant, hepatoprotective, anti-inflammatory, antihyperlipidemic, cytotoxic, anticancer, antioxidant, hypocholesterolemic, antidiabetic, antiallergic, antiviral, antibacterial, antiparasitic, antimicrobial, antifungal, radical scavenging, cardiovascular, wound healing, and detoxification effects [27,60,106,109–111]. Numerous mushroom varieties are recognized for their medicinal properties. For example, *G. lucidum* is often referred to as the 'king of medicinal mushrooms', along with *L. edodes* (shiitake) and *G. frondosa* (maitake), which are widely used for medicinal purposes across many regions of Asia [77,105,112]. In vitro research, in vivo experimentation, and clinical trials involving human subjects have elucidated that mushroom extracts and fresh consumable fungi provide an extensive range of therapeutic advantages, as elaborated upon in the subsequent discussion.

4.1. Anti-Inflammation Property

Inflammation is a defense mechanism in which the blood flow increases to the site of tissue infection, playing a crucial role in the healing process by eliminating harmful cells [113]. However, inflammation also leads to the destruction of cells, which is necessary for recovery. Mushrooms possess properties that allow them to act directly on inflammation. Their lipids, rich in unsaturated fatty acids, exhibit anti-inflammatory qualities as these fatty acids are precursors of eicosanoids involved in balancing inflammatory and anti-inflammatory processes [114]. Mushroom taxa such as *Agaricus* sp., *Pleurotus* sp., and *Termitomyces* sp. exhibit a high abundance of polysaccharides and synthesize biomolecules that play a pivotal role in the protection of joints against inflammatory mechanisms [83,115]. A study was conducted on the mushroom variety *Cordyceps* spp. containing the nucleoside compound cordycepin, which stimulates the generation of interleukin 10; as a result, it is an anti-inflammatory cytokine compound [116]. *H. erinaceus* has also been shown to have anti-inflammatory effects that were demonstrated for both *H. erinaceus* and *H. echinacea*-derived erinacine A, which protect against brain-ischemia-induced neuronal cell death in rats. The mechanism was the suppression of iNOS and MAPK, lowered proinflammatory cytokines, and the mushroom's nerve development capabilities [117].

4.2. Healing Property

Healing is categorized into four stages: hemostasis involving blood clotting, inflammation, proliferation pertaining to tissue growth, and maturation encompassing tissue re-modeling. The repair process is complex and involves various cellular mechanisms such as epithelial cell stimulation, cytokine release, and growth factors. The extract and metabolites from varieties like *G. lucidum* and *A. blazei* (polysaccharides) showed wound-treating properties, including different mechanisms such as epithelial cell stimulation, cytokines, and growth factor release [118]. Chitinous polymers were extracted from the common *A. bisporus* mushrooms by employing straightforward methodologies and subsequently transformed into continuous fibers utilizing a specially designed laboratory-scale fiber-spinning apparatus. The resultant spun fibers consist of an array of chitin fibrils embedded within a glucan matrix, with their fiber dimensions meticulously governed by the specifications associated with needle gauges. After 30 s of contact with a small amount of water (<10 μL), all mushroom chitin fibers demonstrated self-healing characteristics. A microblade may successfully restore macroscopically injured mushroom chitin strands' natural form and tensile characteristics, as indicated by the enhanced self-healing capability for tensile strength (reaching 119%) and breaking strain (attaining 132%). This implies that the process of swelling and deswelling of mushroom chitin fibers may have resulted in the interlocking of chitin fibrils and glucan across the impaired fiber surfaces, resulting in significant self-healing activity [118]. A study was conducted on *G. luciderma* in rats, in which indomethacin caused stomach mucosal lesions, and the polysaccharide fraction induced peptic ulcers for healing in rats [27].

4.3. Enhancing Gut Microflora

Prebiotics are "a substrate that is selectively utilized by host microorganisms conferring a health benefit" [119]. Mushrooms are valuable sources of prebiotics including polyphenols, oligosaccharides, and fibers, which enhance the metabolic activity of beneficial members of gut microflora [117]. A mushroom *G. lucidum* contains polysaccharides and peptides that are non-digestible by pathogens, preventing their multiplication and thereby altering the gut microbiota [11,113]. These indigestible polysaccharides derived from mushrooms serve a prebiotic role, suppressing the proliferation of pathogenic bacteria within the gastrointestinal tract while enhancing the growth of beneficial probiotic bacteria. *G. lucidum*, *H. erinaceus*, *L. edodes*, and *G. frondose* are among the most frequently reported edible mushrooms known to modulate gut flora [119,120]. β-glucan, a type of polysaccharide found in mushrooms, can be fermented by gut bacteria, leading to beneficial changes in the host's microbiome [49]. The diagrammatic representation of how a mushroom-based diet enhances gut microflora is depicted in Figure 2.

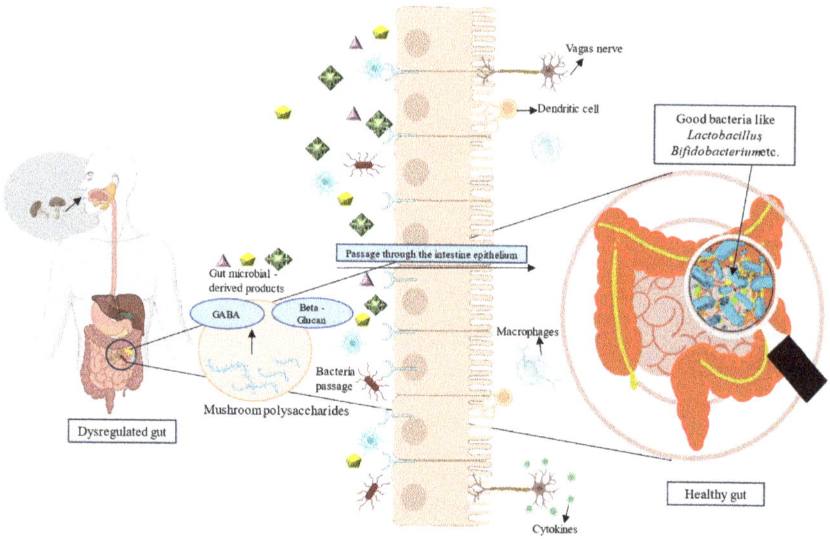

Figure 2. Mushrooms as a potential prebiotic.

4.4. Anticancer Properties

Cancer is a fatal disease causing over 10 million deaths yearly according to the World Health Organization (WHO). Research has demonstrated that polysaccharides derived from mushrooms can inhibit tumor progression by enhancing the immune response, particularly through their impact on natural killer (NK) cells and macrophages via T-cell activation and cytokine secretion [121]. Polysaccharides from mushrooms can impede tumor progression by augmenting the immune response through their influence on natural killer cells and macrophages mediated by T-cell activation and cytokine secretion [122–124]. Notably, nearly 200 species of edible mushrooms demonstrated the capacity to reduce the growth of various cancer cells [125]. Specific compounds found in different mushroom species have been identified for their antitumor properties. For instance, *A. bisporus* contains quinone 490 and 1-oleoyl-2-linoleoyl-3-palmitoyl glycerol, and ganoderiol F and ganodermanontriol in *G. lucidum* and galactoxyloglucan in *H. erinaceous* have shown potential in combating cancer [68]. Mushrooms are rich in various anticancer components such as antroquinonol, krestin, cordycepin, lectin, sulfated polysaccharide hispolon, lentinan, and maitake D fraction [121,126]. The polysaccharide β-glucan is acknowledged for its role in augmenting immune functionality through the stimulation of cytokine synthesis, which subsequently triggers the activation of both phagocytes and leukocytes [80,127]. *L. edodes* contains lentinan and lectins, which demonstrated cytotoxic effects on breast cancer cells [128]. Studies indicate that hispolon, an active polyphenol compound, demonstrates potent antineoplastic effects through multiple mechanisms, including the upregulation of death receptors and downregulation of antiapoptotic proteins like c-FLIP, Bcl-2, and Bcl-xL. Furthermore, hispolon enhances the effectiveness of chemotherapeutic agents, making it a promising candidate for cancer therapy [129]. Furthermore, *G. lucidum* contains certain polysaccharides that are beneficial for mitigating colorectal cancer symptoms as they reduce the expression of rectal cancer-related genes. These polysaccharides also demonstrate cancer-preventive and therapeutic actions by dynamically controlling the gut microbiota and host immune responses. *G. lucidum* polysaccharides can modulate the immune system by activating and expressing cytokines related to inflammation (e.g., interleukin-1, interleukin-6, and tumor necrosis factor-α) and antitumor activity (e.g., interferon-γ and tumor necrosis factor-α). A contemporary in vivo study underscored that a newly identified acid-soluble polysaccharide extracted from *G. frondosa* exhibited protective effects on the thymic and splenic tissues of mice with tumors, concurrently inhibiting the proliferation of H22 solid tumors. These bioactive compounds markedly enhanced the functional activities of natural killer (NK) cells, macrophages, CD19+ B cells, and CD4+ T cells, ultimately facilitating apoptosis of H22 cells through the induction of G0/G1-phase cell cycle arrest [123]. A diagrammatic representation of the mushroom polysaccharides working as anticancer agents is presented in Figure 3.

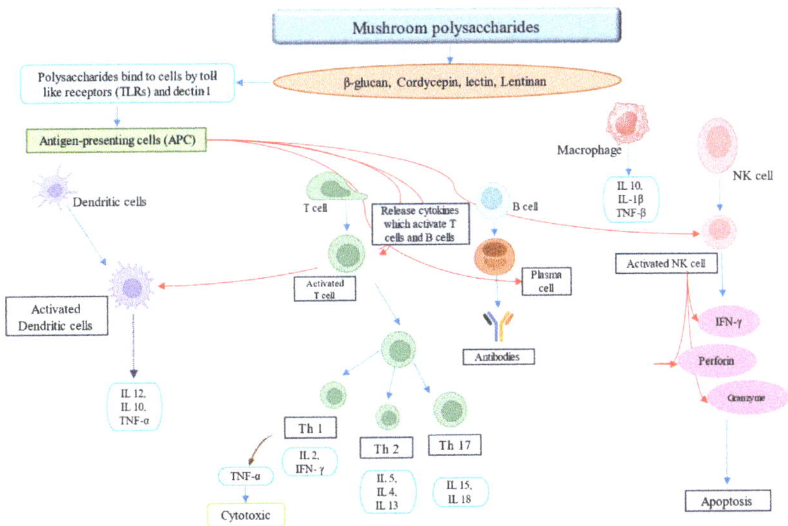

Figure 3. Mushrooms contain certain polysaccharides for the immune response.

4.5. Antioxidant Properties

Oxidative stress can damage DNA, protein, and cell membranes, which eventually leads to various major diseases such as tumors, diabetes, neurodegenerative diseases, and kidney disease [129]. Polysaccharopeptides found in mushrooms can improve overall fitness by triggering enzymes that remove free radicals and reduce oxidative stress [87]. Mushrooms contain a variety of antioxidant compounds including ergothioneine, ergosterol, carotenoids, phenolics, tocopherols (vitamin E), ascorbic acid (vitamin C), polysaccharides (acidic polysaccharides), and amino acids [hydrophobic amino acids (HAAs) like leucine, isoleucine, valine methionine, proline, alanine, etc.] [123,130,131]. For example, *P. ostreatus* extract has been demonstrated to increase catalase gene expression and diminish free radical-induced protein oxidation in adult rats, protecting against age-related illnesses. The ethanolic extract of dietary *P. ostreatus* mushrooms inhibits lipid peroxidation, chelates ferrous ions, reduces ferric ions, and quenches 2,3-diazabicyclo. Another study attributed the superior antioxidant properties of *P. ostreatus* to its carbohydrate component—specifically, β-glucan—which may be responsible for its efficacy [24,132,133]. Furthermore, *P. ostreatus* mushrooms provide a wealth of antioxidants in food sectors, particularly as food additives [134]. An antioxidant assay determined the free radical scavenging activity of *A. bisporus* polysaccharide extracts. At 250 μg/mL, the extract displayed an 86.1% free radical scavenging activity, which was substantially greater ($p < 0.01$) than BHT (83%) [29]. As a result, mushroom consumption may enhance an individual's antioxidative capacity, thereby reducing oxidative stress in the body [119]. The stabilizing of free radicals is shown in Figure 4.

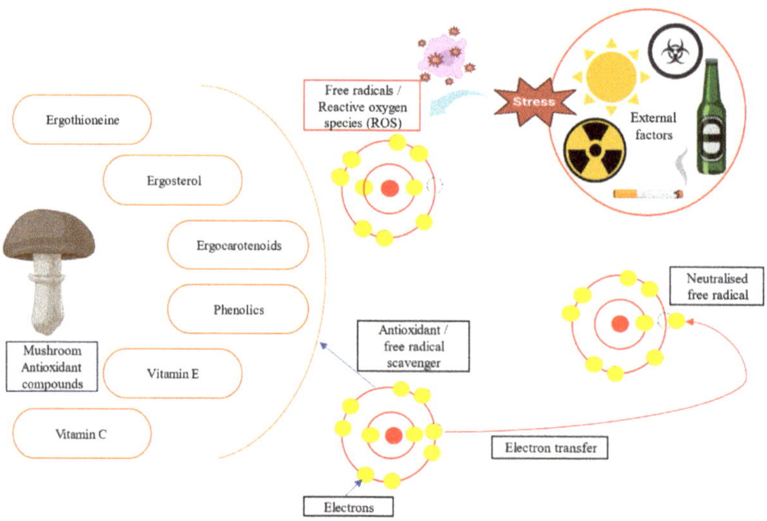

Figure 4. The antioxidant power of mushroom.

4.6. Antidiabetic Properties

Antidiabetic compounds in various mushroom species typically exhibit the following effects: (1) prevention of β cells' apoptosis and promotion of their regeneration; (2) regulation of glucose metabolism; (3) inhibition of inflammation and oxidation; and (4) enhancement of gut microbiota [109,135]. A study on the polysaccharide compounds of *G. lucidum* demonstrated that these compounds reduce insulin resistance without damaging pancreatic islet cells and successfully reverse the process of diabetes [136]. Mushroom extracts from *A. bisporus*, *G. frondosa*, *H. erinaceus*, *G. lucidum*, and *Pleurotus* species reduce blood glucose levels in the liver and muscle by controlling the expression of glycogen synthase kinase (GSK-3β), glycogen synthase (GS), and glucose transporter 4 (GLUT4). As a result, GSK-3β may be identified as a negative regulator that is modulated by insulin-mediated, GS-regulated activity [137]. A study was conducted on high doses of *A. bisporus* extract, which was orally administrated to decrease the severity of streptozotocin-induced hyperglycemia in Sprague–Dawley rats. The rats were provided *A. bisporus* powder (200 mg/kg of body weight) for three weeks, which resulted in a significantly decreased plasma glucose concentration (24.7%), triglyceride content (39.1%), alanine aminotransferase (11.7%), and aspartate aminotransferase (15.7%). Additionally, *G. frondosa* has been noted for its role in blood glucose regulation [138,139].

4.7. Antimicrobial Property

The mushroom species *P. ostreatus* is considered a medicinal mushroom due to its antimicrobial properties because of β-D glucan's presence. It includes several antibacterial agents, such as phenolic compounds, phenolic acids, and flavonoids, which are beneficial in this variety and others [9]. Ethanol extracts from two grown mushroom kinds, *L. edodes* and *A. bisporus*, were tested for antibacterial activity against *Klebsiella pneumoniae*, *Staphylococcus aureus*, *Enterococcus faecalis*, and *Acinetobacter baumannii*. Upon exposure to extracts derived from *L. edodes* and *A. bisporus*, bacterial cell death was observed, attributable to the elevation of protein and DNA levels within the surrounding milieu, indicative of bacterial cell deformation in response to the extracts above. Developing various extracts to combat antibiotic-resistant bacteria is crucial as resistance is anticipated to become one of the most serious health issues in the future. Moreover, there is a significant gap in the literature discussion on the antimicrobial mechanism of mushroom-based compounds [140]. Studies on *P. ostreatus* have demonstrated its effectiveness against Gram-positive bacteria (*Bacillus cereus*, *Bacillus pumilis*, *Micrococcus luteus*, *E. faecalis*, *S. aureus*, and *Bacillus subtilis)* and Gram-negative bacteria (*Klebsiella oxytoca*, *K. pneumonia*, *Shigella* sp., *Salmonella pullorum*, *Salmonella typhi*, *Moraxella* sp., *Escherichia coli*, *Burkholderia pseudomallei*, *Vibrio* sp., and *Pseudomonas aeruginosa*). Moreover, it showed antibacterial action against *Fusarium oxysporum*, *Myrothecium arachidicola*, and *Penicillium rapiricol* [139]. Additionally, *A. bisporus* has demonstrated antibacterial properties against *Neurospora sitophila*, and *Lenzites betulina* has shown antibacterial action against *S. aureus*, *E. coli*, *B. subtilis*, *Fusarium graminearum*, *Gibberella zeae*, and *Cercosporella albo maculans*. *Trichoderma giganteum* has antibacterial action against *F. oxysporum*, *Myrothecium arachidicola*, and *Penicillium rapiricola*. *H. erinaceus* has antibacterial properties against *Helicobacter pylori* [141].

5. Value-Added Products

Several value-added products have been formulated from mushrooms, such as mushroom chips, mushroom soup powder, mushroom pickles, papad, cookies, bhujia, noodles, murabba, yogurt, dried mushrooms, canned mushrooms, mushroom pasta, mushroom kheer, fries, preserve, candies, and mushroom pakora. Additionally, medicinal products such as mushroom pills, mushroom tea, mushroom immunity booster, and protein powder are designed to satisfy taste preferences while providing essential nutrients and bioactive compounds [142,143]. Certain value-added products made from mushrooms are listed in Figure 5.

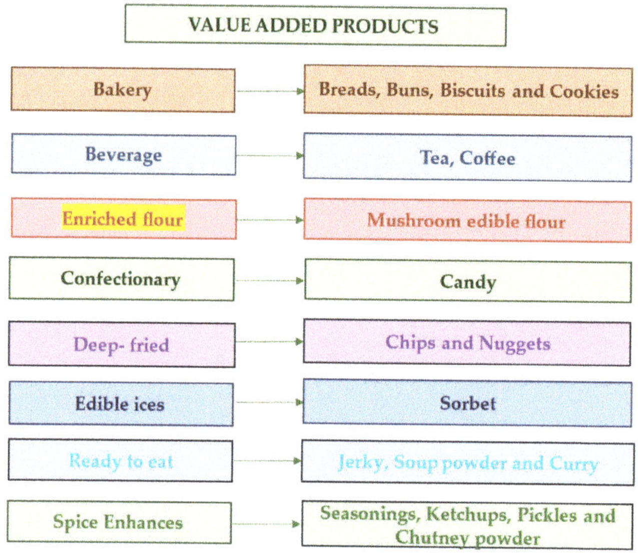

Figure 5. Value-added products prepared from edible mushrooms [105].

Value-added products such as muffins prepared with mushroom powder and white flour have been shown to increase protein, ash, crude fiber, and fat, making them healthier and more nutritious compared to traditional white flour muffins [140]. Functional mushroom cookies and biscuits were also formulated, containing higher nutritional values than those prepared from normal flour. Cookies prepared with *Cordycepes militaris* at concentrations of 1, 3, and 5%, respectively, exhibited increased phenolic and antioxidant contents. Additionally, these cookies showed higher levels of crude fiber, ash content, protein, and crude fat [142]. The incorporation of *C. militar* is flour caused the cookies to become softer, with the hardness slightly decreasing as the concentration of *C. militaris* flour increased ($p > 0.05$). The addition of *C. militaris* flour led to a distorted gluten network, which accounts for the decrease in hardness. Biscuits prepared with a combination of mushroom flour and wheat flour were shown to be more nutritious, with research showing that they can help control diabetes and treat protein–energy shortages. These biscuits also have low GI and GL. Similar to other value-added products, they showed an increase in protein content, ash content, crude fiber, and fat content [143]. The increased properties may be due to bioactive compounds found in wheat and mushroom flours that block alpha-amylase and alpha-glucosidase enzymes. Mushrooms' high fiber content may help with hyperglycemia management. This conclusion is consistent with Owheruo's (2023) discovery that a high-fiber diet lowers blood sugar levels. Nwosu (2022) discovered that those with type 2 diabetes who took a fiber-rich supplement had lower fasting blood glucose levels. According to these findings, mushroom biscuits have the potential to lower blood glucose levels, which may be advantageous for diabetic individuals and those managing hyperglycemia as well as other degenerative illnesses [142,143].

Bars with incorporated dried shiitake mushroom demonstrated hypocholesterolemia and hypoglycemic effects with no toxicity. When assessed for shelf life over 6 months, the bars indicated no significant changes in the microbiological parameters comprising coliforms, *S. aureus*, *B. cereus*, and *Salmonella* sp., with each sample containing fewer than 10 colonies of microorganisms [144]. Similarly, "papad", a trendy dehydrated snack in the Indian market, was fortified with mushroom powder, which increased its protein content, dietary fibers, phosphorous, and calcium [136].

6. Conclusions

Both underdeveloped and developing countries are facing grave issues of malnutrition, poverty, and food insecurity. The consumption and production of highly functional foods, such as mushrooms rich in nutrients and bioactive compounds and offering protection against various diseases, is a step toward address these food issues, as they offer protective and therapeutic benefits against many diseases. The bioactive compounds in mushrooms make them highly suitable for consumption through different sources like food, nutraceuticals, and medicine. Adding mushrooms to our daily diet boosts our nutrient intake by providing essential macro- and micronutrients, and bioactive compounds that are lacking in regular meals. Additionally, this study emphasizes the significant role of mushroom polysaccharides, polyphenols, terpenoids, and glycoproteins in promoting gut health, supporting the immune system, and exhibiting anticancer, antidiabetic, and inflammatory activities. However, further research is needed to elucidate the precise mechanism underlying these health benefits in humans. An in-depth evaluation of mushroom products and varieties in various geographical regions should be performed and technological advancements made for their correct utilization as foods and bioactive agents. Moreover, the production of mushroom-based snacks, beverages, soups, and sauces remains limited on a large scale, and these products often have a short shelf life; therefore, further research is essential to fully understand the potential and limitations of mushroom-based products on the market.

Author Contributions: A.S.: original writing; conceptualization; data curation; formal analysis; funding acquisition; investigation, R.K.S.: reviewing and editing, A.K.: reviewing and editing, P.C.: reviewing and editing, and R.K.: project administration; resources; software; supervision; validation; visualization. All authors have read and agreed to the published version of the manuscript.

Funding: This research received no external funding.

Institutional Review Board Statement: Not applicable.

Informed Consent Statement: Not applicable.

Data Availability Statement: No new data were created or analyzed in this study. Data sharing is not applicable to this article.

Acknowledgments: We share our gratitude to UPES for providing us with all facilities, which included lab facilities, digital library facilities, and access to Science Direct, Google Scholar, PubMed, and all search engines used for writing this review.

Conflicts of Interest: We have no conflicts of interest with anyone.

References

1. Singh, M.P.; Rai, S.N.; Dubey, S.K.; Pandey, A.T.; Tabassum, N.; Chaturvedi, V.K.; Singh, N.B. Biomolecules of mushroom: A recipe of human wellness. *Crit. Rev. Biotechnol.* **2022**, *42*, 913–930. [CrossRef]
2. Liu, S.; Liu, H.; Li, J.; Wang, Y. Research Progress on Elements of Wild Edible Mushrooms. *J. Fungi* **2022**, *8*, 964. [CrossRef] [PubMed]
3. Bhagarathi, L.K.; Subramanian, G.; DaSilva, P.N.B. A review of mushroom cultivation and production, benefits and therapeutic potentials. *World J. Biol. Pharm. Health Sci.* **2023**, *15*, 01–056. [CrossRef]
4. Raut, J.K.; Adhikari, M.K. Mushroom: A true super food. In *Comprehensive Insights in Vegetables of Nepal*; Nepal Academy of Science and Technology (NAST): Lalitpur, Nepal, 2021; pp. 201–234.
5. Kousalya, K.; Krishnakumar, B.; Boomika, S.; Dharati, N.; Hemavathy, N. Edible Mushroom Identification Using Machine Learning. In Proceedings of the 2022 International Conference on Computer Communication and Informatics (ICCCI), Coimbatore, India, 25–27 January 2022.
6. Risoli, S.; Nali, C.; Sarrocco, S.; Cicero, A.F.G.; Colletti, A.; Bosco, F.; Venturella, G.; Gadaleta, A.; Gargano, M.L.; Marcotuli, I. Mushroom-Based Supplements in Italy: Let's Open Pandora's Box. *Nutrients* **2023**, *15*, 776. [CrossRef] [PubMed]

7. Landi, N.; Clemente, A.; Pedone, P.V.; Ragucci, S.; Di Maro, A. An Updated Review of Bioactive Peptides from Mushrooms in a Well-Defined Molecular Weight Range. *Toxins* **2022**, *14*, 84. [CrossRef] [PubMed]
8. Rašeta, M.J.; Rakić, M.S.; Čapelja, E.V.; Karaman, M.A. Update on research data on the nutrient composition of mushrooms and their potentials in future human diets. In *Food Chemistry, Function and Analysis: Edible Fungi: Chemical Composition, Nutrition and Health Effects*; Stojković, D., Barros, L., Eds.; Royal Society of Chemistry: London, UK, 2022; pp. 27–67.
9. Ghosal, A. *Mushroom Cultivation: An Alternate Livelihood Option*; Sasya Shyamala Krishi Vigyan Kendra, Ramakrishna Mission Vivekananda Educational and Research Institute: Arapanch, India, 2023; p. 24.
10. Thakur, M.P. Advances in mushroom production: Key to food, nutritional and employment security: A review. *Indian Phytopathol.* **2020**, *73*, 377–395. [CrossRef]
11. El Sheikha, A.F. Nutritional profile and health benefits of *Ganoderma lucidum* "Lingzhi, Reishi, or Man-nentake" as functional foods: Current scenario and future perspectives. *Foods* **2022**, *11*, 1030. [CrossRef] [PubMed]
12. Suwannarach, N.; Kumla, J.; Zhao, Y.; Kakumyan, P. Impact of Cultivation Substrate and Microbial Community on Improving Mushroom Productivity: A Review. *Biology* **2022**, *11*, 569. [CrossRef] [PubMed]
13. Mykchaylova, O.B.; Poyedinok, N.L.; Shchetinin, V.M. Screening of strains of the medicinal mushroom *Fomitopsis officinalis* (Vill.) Bondartsev & Singer promising for biotechnological use. Innovative Biosyst. *Bioeng.* **2022**, *6*, 110–118.
14. Zeng, X.; Li, J.; Lyu, X.; Chen, T.; Chen, J.; Chen, X.; Guo, S. Utilization of functional agro-waste residues for oyster mushroom production: Nutritions and active ingredients in healthcare. *Front. Plant Sci.* **2023**, *13*, 1085022. [CrossRef] [PubMed]
15. Kumla, J.; Suwannarach, N.; Sujarit, K.; Penkhrue, W.; Kakumyan, P.; Jatuwong, K.; Vadthanarat, S.; Lumyong, S. Cultivation of mushrooms and their lignocellulolytic enzyme production through the utilization of agro-industrial waste. *Molecules* **2020**, *25*, 2811. [CrossRef] [PubMed]
16. Zied, D.C.; Sánchez, J.E.; Noble, R.; Pardo-Giménez, A. Use of spent mushroom substrate in new mushroom crops to promote the transition towards a circular economy. *Agronomy* **2020**, *10*, 1239. [CrossRef]
17. Ahmed, R.; Niloy, M.A.H.M.; Islam, M.S.; Reza, M.S.; Yesmin, S.; Rasul, S.B.; Khandakar, J. Optimizing tea waste as a sustainable substrate for oyster mushroom (*Pleurotus ostreatus*) cultivation: A comprehensive study on biological efficiency and nutritional aspect. *Front. Sustain. Food Syst.* **2024**, *7*, 1308053. [CrossRef]
18. Chen, L.; Qian, L.; Zhang, X.; Li, J.; Zhang, Z.; Chen, X. Research progress on indoor environment of mushroom factory. *Int. J. Agric. Biol. Eng.* **2022**, *15*, 25–32. [CrossRef]
19. Krupodorova, T.A.; Barshteyn, V.Y.; Sekan, A.S. Review of the basic cultivation conditions influence on the growth of basidiomycetes. *Curr. Res. Environ. Appl. Mycol.* **2021**, *11*, 494–531. [CrossRef]
20. Martinez-Medina, G.A.; Chávez-González, M.L.; Verma, D.K.; Prado-Barragán, L.A.; Martínez-Hernández, J.L.; Flores-Gallegos, A.C.; Thakur, M.; Srivastav, P.P.; Aguilar, C.N. Bio-functional components in mushrooms, a health opportunity: Ergothionine and huitlacohe as recent trends. *J. Funct. Foods* **2021**, *77*, 104326. [CrossRef]
21. Hu, Y.; Mortimer, P.E.; Hyde, K.D.; Kakumyan, P.; Thongklang, N. Mushroom cultivation for soil amendment and bioremediation. *Circ. Agric. Syst.* **2021**, *1*, 11. [CrossRef]
22. Raman, J.; Jang, K.Y.; Oh, Y.L.; Oh, M.; Im, J.H.; Lakshmanan, H.; Sabaratnam, V. Cultivation and Nutritional Value of Prominent *Pleurotus* spp.: An Overview. *Mycobiology* **2021**, *49*, 1–14. [CrossRef]
23. Arunachalam, K.; Sreeja, P.S.; Yang, X. The Antioxidant Properties of Mushroom Polysaccharides can Potentially Mitigate Oxidative Stress, Beta-Cell Dysfunction and Insulin Resistance. *Front. Pharmacol.* **2022**, *13*, 874474. [CrossRef]
24. Stastny, J.; Marsik, P.; Tauchen, J.; Bozik, M.; Mascellani, A.; Havlik, J.; Landa, P.; Jablonsky, I.; Treml, J.; Herczogova, P.; et al. Antioxidant and Anti-Inflammatory Activity of Five Medicinal Mushrooms of the Genus Pleurotus. *Antioxidants* **2022**, *11*, 1569. [CrossRef]
25. Araújo-Rodrigues, H.; Sousa, A.S.; Relvas, J.B.; Tavaria, F.K.; Pintado, M. An overview on mushroom polysaccharides: Health-promoting properties, prebiotic and gut microbiota modulation effects and structure-function correlation. *Carbohydr. Polym.* **2024**, *333*, 121978. [CrossRef] [PubMed]
26. Bains, A.; Chawla, P.; Kaur, S.; Najda, A.; Fogarasi, M.; Fogarasi, S. Bioactives from mushroom: Health attributes and food industry applications. *Materials* **2021**, *14*, 7640. [CrossRef]
27. Chopra, H.; Mishra, A.K.; Baig, A.A.; Mohanta, T.K.; Mohanta, Y.K.; Baek, K.H. Narrative review: Bio-active potential of various mushrooms as the treasure of versatile therapeutic natural product. *J. Fungi* **2021**, *7*, 728. [CrossRef] [PubMed]
28. Sachdeva, V.; Roy, A.; Bharadvaja, N. Current Prospects of Nutraceuticals: A Review. *Curr. Pharm. Biotechnol.* **2020**, *21*, 884–896. [CrossRef]
29. Dimopoulou, M.; Kolonas, A.; Mourtakos, S.; Androutsos, O.; Gortzi, O. Nutritional composition and biological properties of sixteen edible mushroom species. *Appl. Sci.* **2022**, *12*, 8074. [CrossRef]
30. Shbeeb, D.A.; Farahat, M.F.; Ismail, H.M. Macronutrients analysis of fresh and canned *Agaricus bisporus* and *Pleurotus ostreatus* mushroom species sold in Alexandria markets, Egypt. *Egypt. Prog. Nutr* **2019**, *21*, 203–209.

31. Bandara, A.R.; Rapior, S.; Mortimer, P.E.; Kakumyan, P.; Hyde, K.D.; Xu, J. A review of the polysaccharide, protein and selected nutrient content of Auricularia, and their potential pharmacological value. *Mycosphere J.* **2019**, *10*, 579–607. [CrossRef]
32. Rahman, M.A.; Masud, A.A.; Lira, N.Y.; Shakil, S. Proximate analysis, phytochemical screening and antioxidant activity of different strains of Auricularia auricula-judae (ear mushroom). *Int. J. Tradit. Complement. Med.* **2020**, *5*, 29.
33. Sangthong, S.; Pintathong, P.; Pongsua, P.; Jirarat, A.; Chaiwut, P. Polysaccharides from Volvariella volvacea mushroom: Extraction, biological activities and cosmetic efficacy. *J. Fungi* **2022**, *8*, 572. [CrossRef] [PubMed]
34. Subbiah, K.A.; Balan, V. A comprehensive review of tropical milky white mushroom (Calocybe indica P&C). *Mycobiology* **2015**, *43*, 184–194. [PubMed]
35. Mišković, J.; Rašeta, M.; Krsmanović, N.; Karaman, M. Update on mycochemical profile and selected biological activities of genus Schizophyllum Fr. 1815. *Microbiol. Res.* **2023**, *14*, 409–429. [CrossRef]
36. Gharib, M.A.A.; Elhassaneen, Y.A.E.E.; Radwan, H. Nutrients and nutraceuticals content and in vitro biological activities of reishi mushroom (*Ganoderma lucidum*) fruiting bodies. *Alex. Sci. Exch. J.* **2022**, *43*, 301–316. [CrossRef]
37. Nwe, M.L.; Zin, T.T. Phytochemistry and pharmacological studies on *Flammulina velutipes* (Curtis). *Int. J. Sci. Res. Eng. Dev.* **2020**, *3*, 32–37.
38. Wu, J.Y.; Siu, K.C.; Geng, P. Bioactive ingredients and medicinal values of *Grifola frondosa* (Maitake). *Foods* **2021**, *10*, 95. [CrossRef]
39. Ponnusamy, C.; Uddandrao, V.V.S.; Pudhupalayam, S.P.; Singaravel, S.; Periyasamy, T.; Ponnusamy, P.; Prabhu, P.; Sasikumar, V.; Ganapathy, S. *Lentinula edodes* (edible mushroom) as a nutraceutical: A review. *Biosci. Biotechnol. Res. Asia* **2022**, *19*, 1–11. [CrossRef]
40. Łysakowska, P.; Sobota, A.; Wirkijowska, A. Medicinal mushrooms: Their bioactive components, nutritional value and application in functional food production—A review. *Molecules* **2023**, *28*, 5393. [CrossRef]
41. Leong, Y.K.; Yang, F.C.; Chang, J.S. Extraction of polysaccharides from edible mushrooms: Emerging technologies and recent advances. *Carbohydr. Polym.* **2021**, *251*, 117006. [CrossRef] [PubMed]
42. Seo, D.J.; Choi, C. Antiviral bioactive compounds of mushrooms and their antiviral mechanisms: A review. *Viruses* **2021**, *13*, 350. [CrossRef] [PubMed]
43. Yu, C.; Dong, Q.; Chen, M.; Zhao, R.; Zha, L.; Zhao, Y.; Zhang, M.; Zhang, B.; Ma, A. The effect of mushroom dietary fiber on the gut microbiota and related health benefits: A review. *J. Fungi* **2023**, *9*, 1028. [CrossRef]
44. Asadpoor, M.; Ithakisiou, G.N.; Henricks, P.A.; Pieters, R.; Folkerts, G.; Braber, S. Non-digestible oligosaccharides and short chain fatty acids as therapeutic targets against enterotoxin-producing bacteria and their toxins. *Toxins* **2021**, *13*, 175. [CrossRef] [PubMed]
45. Asadpoor, M.; Peeters, C.; Henricks, P.A.; Varasteh, S.; Pieters, R.J.; Folkerts, G.; Braber, S. Anti-pathogenic functions of non-digestible oligosaccharides in vitro. *Nutrients* **2020**, *12*, 1789. [CrossRef] [PubMed]
46. Yang, S.; Wu, C.; Yan, Q.; Li, X.; Jiang, Z. Nondigestible functional oligosaccharides: Enzymatic production and food applications for intestinal health. *Annu. Rev. Food Sci. Technol.* **2023**, *14*, 297–322. [CrossRef] [PubMed]
47. Pandey, M.; Satisha, G.C.; Azeez, S.; Kumaran, G.S.; Chandrashekara, C. Mushrooms for integrated and diversified nutrition. *J. Hortic. Sci.* **2022**, *17*, 6–18. [CrossRef]
48. Antunes, F.; Marçal, S.; Taofiq, O.; Morais, A.M.M.B.; Freitas, A.C.; Ferreira, I.C.F.R.; Pintado, M. Valorization of mushroom by-products as a source of value-added compounds and potential applications. *Molecules* **2020**, *25*, 2672. [CrossRef] [PubMed]
49. Chugh, R.M.; Mittal, P.; Mp, N.; Arora, T.; Bhattacharya, T.; Chopra, H.; Cavalu, S.; Gautam, R.K. Fungal mushrooms: A natural compound with therapeutic applications. *Front. Pharmacol.* **2022**, *13*, 925387. [CrossRef]
50. Assemie, A.; Abaya, G. The Effect of Edible Mushroom on Health and Their Biochemistry. *Int. J. Microbiol.* **2022**, *2022*, 8744788. [CrossRef]
51. Pathak, M.P.; Pathak, K.; Saikia, R.; Gogoi, U.; Ahmad, M.Z.; Patowary, P.; Das, A. Immunomodulatory effect of mushrooms and their bioactive compounds in cancer: A comprehensive review. *Biomed. Pharmacother.* **2022**, *149*, 112901. [CrossRef] [PubMed]
52. Singh, R.S.; Walia, A.K.; Kennedy, J.F. Mushroom lectins in biomedical research and development. *Int. J. Biol. Macromol.* **2020**, *151*, 1340–1350. [CrossRef] [PubMed]
53. Łusarczyk, J.; Adamska, E.; Czerwik-Marcinkowska, J. Fungi and algae as sources of medicinal and other biologically active compounds: A review. *Nutrients* **2021**, *13*, 3178. [CrossRef]
54. Ayimbila, F.; Keawsompong, S. Nutritional Quality and Biological Application of Mushroom Protein as a Novel Protein Alternative. *Curr. Nutr. Rep.* **2023**, *12*, 290–307. [CrossRef]
55. Päivärinta, E.; Itkonen, S.T.; Pellinen, T.; Lehtovirta, M.; Erkkola, M.; Pajari, A.M. Replacing animal-based proteins with plant-based proteins changes the composition of a whole Nordic diet—A randomised clinical trial in healthy Finnish adults. *Nutrients* **2020**, *12*, 943. [CrossRef] [PubMed]
56. Gorissen, S.H.; Crombag, J.J.; Senden, J.M.; Waterval, W.H.; Bierau, J.; Verdijk, L.B.; van Loon, L.J. Protein content and amino acid composition of commercially available plant-based protein isolates. *Amino Acids* **2018**, *50*, 1685–1695. [CrossRef] [PubMed]

57. Langyan, S.; Yadava, P.; Khan, F.N.; Dar, Z.A.; Singh, R.; Kumar, A. Sustaining protein nutrition through plant-based foods. *Front. Nutr.* **2022**, *8*, 772573. [CrossRef] [PubMed]
58. Qiu, Y.; Lin, G.; Liu, W.; Zhang, F.; Linhardt, R.J.; Wang, X.; Zhang, A. Bioactive compounds in *Hericium erinaceus* and their biological properties: A review. *Food Sci. Hum. Wellness* **2024**, *13*, 1825–1844. [CrossRef]
59. Kostanda, E.; Musa, S.; Pereman, I. Unveiling the Chemical Composition and Biofunctionality of *Hericium* spp. Fungi: A Comprehensive Overview. *Int. J. Mol. Sci.* **2024**, *25*, 5949. [CrossRef]
60. Venturella, G.; Ferraro, V.; Cirlincione, F.; Gargano, M.L. Medicinal mushrooms: Bioactive compounds, use, and clinical trials. *Int. J. Mol. Sci.* **2021**, *22*, 634. [CrossRef]
61. Gu, H.; Liang, L.; Kang, Y.; Yu, R.; Wang, J.; Fan, D. Preparation, characterization, and property evaluation of *Hericium erinaceus* peptide–calcium chelate. *Front. Nutr.* **2024**, *10*, 1337407. [CrossRef]
62. Jovanović, J.A.; Mihailović, M.; Usković, A.; Grdović, N.; Dinić, S.; Vidaković, M. The effects of major mushroom bioactive compounds on mechanisms that control blood glucose level. *J. Fungi* **2021**, *7*, 58. [CrossRef] [PubMed]
63. Liu, X.; Huang, Y.; Wang, J.; Zhou, S.; Wang, Y.; Cai, M.; Yu, L. A study on the antioxidant properties and stability of ergothioneine from culinary-medicinal mushrooms. *Int. J. Med. Mushrooms* **2020**, *22*, 211–220. [CrossRef] [PubMed]
64. Kalaras, M.D.; Richie, J.P.; Calcagnotto, A.; Beelman, R.B. Mushrooms: A rich source of the antioxidants ergothioneine and glutathione. *Food Chem.* **2017**, *233*, 429–433. [CrossRef] [PubMed]
65. Lam-Sidun, D.; Peters, K.M.; Borradaile, N.M. Mushroom-derived medicine? Preclinical studies suggest potential benefits of ergothioneine for cardiometabolic health. *Int. J. Mol. Sci.* **2021**, *22*, 3246. [CrossRef]
66. Amaranthus, M. A Mushroom-derived Compound that Could Change your Life: Ergothioneine. *Food Nutr. J.* **2023**, *8*, 282.
67. Sinha, S.K.; Upadhyay, T.K.; Sharma, S.K. Nutritional-medicinal profile and quality categorization of fresh white button mushroom. Biointerface Res. *Appl. Chem* **2021**, *11*, 8669–8685.
68. Saini, R.K.; Rauf, A.; Khalil, A.A.; Ko, E.-Y.; Keum, Y.-S.; Anwar, S.; Alamri, A.; Rengasamy, K.R. Edible mushrooms show significant differences in sterols and fatty acid compositions. *S. Afr. J. Bot.* **2021**, *141*, 344–356. [CrossRef]
69. Rangel-Vargas, E.; Rodriguez, J.A.; Domínguez, R.; Lorenzo, J.M.; Sosa, M.E.; Andrés, S.C.; Rosmini, M.; Pérez-Alvarez, J.A.; Teixeira, A.; Santos, E.M. Edible mushrooms as a natural source of food ingredient/additive replacer. *Foods* **2021**, *10*, 2687. [CrossRef]
70. Awfa, A.; Amany, K.; Fahmida, K.; Farida, K.; Saud, A.; Sohair, S.; Somaia, I.; Mohammed, E.E.; Ahmed, G.; Saud, A.; et al. Nutrition Value of Mushroom Intake And Its Impact On Human Health. *Int. Neurourol. J.* **2024**, *27*, 12–17.
71. Abdelshafy, A.M.; Belwal, T.; Liang, Z.; Wang, L.; Li, D.; Luo, Z.; Li, L. A comprehensive review on phenolic compounds from edible mushrooms: Occurrence, biological activity, application and future prospective. *Crit. Rev. Food Sci. Nutr.* **2022**, *62*, 6204–6224. [CrossRef]
72. Cateni, F.; Gargano, M.L.; Procida, G.; Venturella, G.; Cirlincione, F.; Ferraro, V. Mycochemicals in wild and cultivated mushrooms: Nutrition and health. *Phytochem. Rev.* **2022**, *21*, 339–383. [CrossRef]
73. Niazi, A.R.; Ghafoor, A. Different ways to exploit mushrooms: A review. *All Life* **2021**, *14*, 450–460. [CrossRef]
74. Sharif, S.; Mustafa, G.; Munir, H.; Weaver, C.M.; Jamil, Y.; Shahid, M. Proximate composition and micro-nutrient mineral profile of wild *Ganoderma lucidum* and four commercial exotic mushrooms by ICP-OES and LIBS. *J. Food Nutr. Res.* **2016**, *4*, 703–708.
75. Dawadi, E.; Magar, P.B.; Bhandari, S.; Subedi, S.; Shrestha, S.; Shrestha, J. Nutritional and post-harvest quality preservation of mushrooms: A review. *Heliyon* **2022**, *8*, e12093. [CrossRef] [PubMed]
76. Motta, F.; Gershwin, M.E.; Selmi, C. Mushrooms and immunity. *J. Autoimmun.* **2021**, *117*, 102576. [CrossRef] [PubMed]
77. Zhou, Z.; Liang, S.; Zou, X.; Teng, Y.; Wang, W.; Fu, L. Determination of Phenolic Acids Using Ul-tra-High-Performance Liquid Chromatography Coupled with Triple Quadrupole (UHPLC-QqQ) in Fruiting Bodies of *Sanghuangporus baumii* (Pilát) LW Zhou and YC Dai. *Plants* **2023**, *12*, 3565. [CrossRef]
78. Sova, M.; Saso, L. Natural sources, pharmacokinetics, biological activities and health benefits of hy-droxycinnamic acids and their metabolites. *Nutrients* **2020**, *12*, 2190. [CrossRef] [PubMed]
79. Naim, M.J. A review on mushrooms as a versatile therapeutic agent with emphasis on its bioactive con-stituents for anticancer and antioxidant potential. *Explor. Med.* **2024**, *5*, 312–330. [CrossRef]
80. Park, H.J. Current uses of mushrooms in cancer treatment and their anticancer mechanisms. *Int. J. Mol. Sci.* **2022**, *23*, 10502. [CrossRef]
81. Zhu, F.; Zhang, Q.; Feng, J.; Zhang, X.; Li, T.; Liu, S.; Chen, Y.; Li, X.; Wu, Q.; Xue, Y.; et al. β-Glucan produced by Lentinus edodes suppresses breast cancer progression via the inhibition of macrophage M2 polarization by integrating au-tophagy and inflammatory signals. *Immun. Inflamm. Dis.* **2023**, *11*, e876. [CrossRef]
82. Ahmed Elkhateeb, W.; Mosbah Daba, G. Medicinal mushroom: What should we know? *Int. J. Pharm. Chem. Anal.* **2022**, *9*, 1–9. [CrossRef]
83. Rani, M.; Mondal, S.M.; Kundu, P.; Thakur, A.; Chaudhary, A.; Vashistt, J.; Shankar, J. Edible mushroom: Occurrence, management and health benefits. *Food Mater. Res.* **2023**, *3*, 21. [CrossRef]

84. Sławińska, A.; Sołowiej, B.G.; Radzki, W.; Fornal, E. Wheat Bread Supplemented with *Agaricus bisporus* Powder: Effect on Bioactive Substances Content and Technological Quality. *Foods* **2022**, *11*, 3786. [CrossRef] [PubMed]
85. Rowaiye, A.; Wilfred, O.I.; Onuh, O.A.; Bur, D.; Oni, S.; Nwonu, E.J.; Ibeanu, G.; Oli, A.N.; Wood, T.T. Modulatory Effects of Mushrooms on the Inflammatory Signaling Pathways and Pro-inflammatory Mediators. *Clin. Complement. Med. Pharmacol.* **2022**, *2*, 100037. [CrossRef]
86. Elhusseiny, S.M.; El-Mahdy, T.S.; Elleboudy, N.S.; Farag, M.M.S.; Aboshanab, K.M.; Yassien, M.A. Immunomodulatory activity of extracts from five edible basidiomycetes mushrooms in Wistar albino rats. *Sci. Rep.* **2022**, *12*, 12423. [CrossRef]
87. Fang, D.; Wang, D.; Ma, G.; Ji, Y.; Zheng, H.; Chen, H.; Zhao, M.; Hu, Q.; Zhao, L. Auricularia polytricha noodles prevent hyperlipemia and modulate gut microbiota in high-fat diet fed mice. *Food Sci. Hum. Wellness* **2021**, *10*, 431–441. [CrossRef]
88. Miao, J.; Regenstein, J.M.; Qiu, J.; Zhang, J.; Zhang, X.; Li, H.; Zhang, H.; Wang, Z. Isolation, structural characterization and bioactivities of polysaccharides and its derivatives from Auricularia-A review. *Int. J. Biol. Macromol.* **2020**, *150*, 102–113. [CrossRef] [PubMed]
89. Su, Y.; Li, L. Structural characterization and antioxidant activity of polysaccharide from four auriculariales. *Carbohydr. Polym.* **2020**, *229*, 115407. [CrossRef] [PubMed]
90. Ghosh, K. A Review on Edible Straw Mushrooms: A Source of High Nutritional Supplement, Biologically Active Diverse Structural Polysaccharides. *J. Sci. Res.* **2020**, *64*, 295–304. [CrossRef]
91. Chelladurai, G.; Yadav, T.K.; Pathak, R.K. Chemical composition and nutritional value of paddy straw milky mushroom (*Calocybe indica*). *Nat. Environ. Pollut. Technol.* **2021**, *20*, 1157–1164. [CrossRef]
92. Shashikant, M.; Bains, A.; Chawla, P.; Fogarasi, M.; Fogarasi, S. The Current Status, Bioactivity, Food, and Pharmaceutical Approaches of *Calocybe indica*: A Review. *Antioxidants* **2022**, *11*, 1145. [CrossRef]
93. Chen, Z.; Yin, C.; Fan, X.; Ma, K.; Yao, F.; Zhou, R.; Shi, D.; Cheng, W.; Gao, H. Characterization of physicochemical and biological properties of *Schizophyllum commune* polysaccharide extracted with different methods. *Int. J. Biol. Macromol.* **2020**, *156*, 1425–1434. [CrossRef]
94. Wongaem, A.; Reamtong, O.; Srimongkol, P.; Sangtanoo, P.; Saisavoey, T.; Karnchanatat, A. Antioxidant properties of peptides obtained from the split gill mushroom (*Schizophyllum commune*). *J. Food Sci. Technol.* **2021**, *58*, 680–691. [CrossRef]
95. Ahmad, R.; Riaz, M.; Khan, A.; Aljamea, A.; Algheryafi, M.; Sewaket, D.; Alqathama, A. *Ganoderma lucidum* (Reishi) an edible mushroom; a comprehensive and critical review of its nutritional, cosmeceutical, mycochemical, pharmacological, clinical, and toxicological properties. *Phytother. Res.* **2021**, *35*, 6030–6062. [CrossRef] [PubMed]
96. Pattanayak, S.; Das, S.; Biswal, G. Ganoderma: The wild mushroom with wonderful health benefits. *J. Pharmacogn. Phytochem.* **2020**, *9*, 313–316. [CrossRef]
97. Aursuwanna, T.; Noitang, S.; Sangtanoo, P.; Srimongkol, P.; Saisavoey, T.; Puthong, S.; Reamtong, O.; Karnchanatat, A. Investigating the cellular antioxidant and anti-inflammatory effects of the novel peptides in lingzhi mushrooms. *Heliyon* **2022**, *8*, e11067. [CrossRef] [PubMed]
98. Cör Andrejč, D.; Knez, Ž.; Knez Marevci, M. Antioxidant, antibacterial, antitumor, antifungal, antiviral, anti-inflammatory, and nevro-protective activity of *Ganoderma lucidum*: An overview. *Front. Pharmacol.* **2022**, *13*, 934982. [CrossRef] [PubMed]
99. Li, C.; Wu, G.; Zhao, H.; Dong, N.; Wu, B.; Chen, Y.; Lu, Q. Natural-Derived Polysaccharides From Plants, Mushrooms, and Seaweeds for the Treatment of Inflammatory Bowel Disease. *Front. Pharmacol.* **2021**, *12*, 651813. [CrossRef]
100. Desisa, B.; Muleta, D.; Jida, M.; Dejene, T.; Goshu, A.; Negi, T.; Martin-Pinto, P. Improvement of nutritional composition of shiitake mushroom (*Lentinula edodes*) using formulated substrates of plant and animal origins. *Future Foods* **2024**, *9*, 100302. [CrossRef]
101. Banerjee, D.K.; Das, A.K.; Banerjee, R.; Pateiro, M.; Nanda, P.K.; Gadekar, Y.P.; Biswas, S.; McClements, D.J.; Lorenzo, J.M. Application of enoki mushroom (*Flammulina velutipes*) stem wastes as functional ingredients in goat meat nuggets. *Foods* **2020**, *9*, 432. [CrossRef] [PubMed]
102. Das, A.K.; Nanda, P.K.; Dandapat, P.; Bandyopadhyay, S.; Gullón, P.; Sivaraman, G.K.; McClements, D.J.; Gullón, B.; Lorenzo, J.M. Edible mushrooms as functional ingredients for development of healthier and more sustainable muscle foods: A flexitarian approach. *Molecules* **2021**, *26*, 2463. [CrossRef]
103. González-Quero, N.; Martínez, P. Bioactive compounds in some principal mushrooms: An association to adverse effects. *GSC Adv. Res. Rev.* **2020**, *5*, 031–047. [CrossRef]
104. Tachabenjarong, N.; Rungsardthong, V.; Ruktanonchi, U.; Poodchakarn, S.; Thumthanaruk, B.; Vatanyoopaisarn, S.; Suttisintong, K.; Iempridee, T.; Uttapap, D. Bioactive compounds and antioxidant activity of Lion's Mane mushroom (*Hericium erinaceus*) from different growth periods. *E3S Web Conf.* **2022**, *355*, 02016. [CrossRef]
105. Yadav, D.; Negi, P.S. Bioactive components of mushrooms: Processing effects and health benefits. *Food Res. Int.* **2021**, *148*, 110599. [CrossRef] [PubMed]
106. Li, M.; Yu, L.; Zhao, J.; Zhang, H.; Chen, W.; Zhai, Q.; Tian, F. Role of dietary edible mushrooms in the modulation of gut microbiota. *J. Funct. Foods* **2021**, *83*, 104538. [CrossRef]

107. Singh, R.P.; Bhardwaj, A. β-glucans: A potential source for maintaining gut microbiota and the immune system. *Front. Nutr.* **2023**, *10*, 1143682. [CrossRef]
108. Murphy, E.J.; Rezoagli, E.; Major, I.; Rowan, N.J.; Laffey, J.G. β-glucan metabolic and immunomodulatory properties and potential for clinical application. *J. Fungi* **2020**, *6*, 356. [CrossRef] [PubMed]
109. Elkhateeb, W.A. What Medicinal Mushroom Can Do? *Chem. Res. J.* **2020**, *5*, 106–118.
110. Reis, F.S.; Hu, Y.; Fernandes, T.H. Mushrooms as future generation healthy foods. *Front. Nutr.* **2022**, *9*, 1050099.
111. Cardoso, R.V.; Oludemi, T.; Fernandes, Â.; Ferreira, I.C.; Barros, L. *Bioactive Properties of Mushrooms with Potential Health Benefits*; The Royal Society of Chemistry: London, UK, 2022.
112. Yin, C.; Noratto, G.D.; Fan, X.; Chen, Z.; Yao, F.; Shi, D.; Gao, H. The Impact of Mushroom Polysaccharides on Gut Microbiota and Its Beneficial Effects to Host: A Review. *Carbohydr. Polym.* **2020**, *250*, 116942. [CrossRef]
113. Ambhore, J.P.; Adhao, V.S.; Rafique, S.S.; Telgote, A.A.; Dhoran, R.S.; Shende, B.A. A concise review: Edible mushroom and their medicinal significance. *Explor. Foods Foodomics* **2024**, *2*, 183–194. [CrossRef]
114. Bhambri, A.; Srivastava, M.; Mahale, V.G.; Mahale, S.; Karn, S.K. Mushrooms as Potential Sources of Active Metabolites and Medicines. *Front. Microbiol.* **2022**, *13*, 837266. [CrossRef]
115. Yin, Z.; Zhang, J.; Qin, J.; Guo, L.; Guo, Q.; Kang, W.; Ma, C.; Chen, L. Anti-inflammatory properties of poly-saccharides from edible fungi on health-promotion: A review. *Front. Pharmacol.* **2024**, *15*, 1447677. [CrossRef] [PubMed]
116. Hetland, G.; Tangen, J.M.; Mahmood, F.; Mirlashari, M.R.; Nissen-Meyer, L.S.; Nentwich, I.; Therkelsen, S.P.; Tjønnfjord, G.E.; Johnson, E. Antitumor, anti-inflammatory and antiallergic effects of Agaricus blazei mushroom extract and the related medicinal basidiomycetes mushrooms, *Hericium erinaceus* and *Grifola frondosa*: A review of preclinical and clinical studies. *Nutrients* **2020**, *12*, 1339. [CrossRef] [PubMed]
117. Sharifi-Rad, J.; Butnariu, M.; Ezzat, S.M.; Adetunji, C.O.; Imran, M.; Sobhani, S.R.; Tufail, T.; Hosseinabadi, T.; Ramírez-Alarcón, K.; Martorell, M.; et al. Mushrooms-Rich Preparations on Wound Healing: From Nutritional to Medicinal Attributes. *Front. Pharmacol.* **2020**, *11*, 567518. [CrossRef] [PubMed]
118. Lee, J.; Park, Y. Self-Healing Properties of Fibers Constructed from Mushroom-Derived Chitinous Polymers. *ACS Sustain. Chem. Eng.* **2023**, *11*, 2959–2967. [CrossRef]
119. Zhao, J.; Hu, Y.; Qian, C.; Hussain, M.; Liu, S.; Zhang, A.; He, R.; Sun, P. The Interaction between Mushroom Polysaccharides and Gut Microbiota and Their Effect on Human Health: A Review. *Biology* **2023**, *12*, 122. [CrossRef]
120. Sivanesan, I.; Muthu, M.; Gopal, J.; Oh, J.W. Mushroom polysaccharide-assisted anticarcinogenic myco-therapy: Reviewing its clinical trials. *Molecules* **2022**, *27*, 4090. [CrossRef] [PubMed]
121. Fan, J.; Zhu, J.; Zhu, H.; Zhang, Y.; Xu, H. Potential therapeutic target for polysaccharide inhibition of colon cancer progression. *Front. Med.* **2024**, *10*, 1325491. [CrossRef] [PubMed]
122. Paludan, S.R.; Pradeu, T.; Masters, S.L.; Mogensen, T.H. Constitutive immune mechanisms: Mediators of host defence and immune regulation. *Nat. Rev. Immunol.* **2021**, *21*, 137–150. [CrossRef] [PubMed]
123. Li, W.; Zhou, Q.; Lv, B.; Li, N.; Bian, X.; Chen, L.; Kong, M.; Shen, Y.; Zheng, W.; Zhang, J.; et al. *Ganoderma lucidum* Polysaccharide Supplementation Significantly Activates T-Cell-Mediated Antitumor Immunity and Enhances Anti-PD-1 Immunotherapy Efficacy in Colorectal Cancer. *J. Agric. Food Chem.* **2024**, *72*, 12072–12082. [CrossRef]
124. Hilszczańska, D. Healing Properties of Edible Mushrooms. In *Advances in Macrofungi*; CRC Press: Boca Raton, FL, USA, 2021; pp. 39–51.
125. Liu, M.M.; Liu, T.; Yeung, S.; Wang, Z.; Andresen, B.; Parsa, C.; Orlando, R.; Zhou, B.; Wu, W.; Li, X.; et al. Inhibitory activity of medicinal mushroom *Ganoderma lucidum* on colorectal cancer by attenuating inflammation. *Precis. Clin. Med.* **2021**, *4*, 231–245. [CrossRef]
126. Zhong, Y.; Tan, P.; Lin, H.; Zhang, D.; Chen, X.; Pang, J.; Mu, R. A review of *Ganoderma lucidum* polysaccharide: Preparations, structures, physicochemical properties and application. *Foods* **2024**, *13*, 2665. [CrossRef]
127. Xu, H.; Zou, S.; Xu, X. The β-glucan from Lentinula edodes suppresses cell proliferation and promotes apoptosis in estrogen receptor positive breast cancers. *Oncotarget* **2017**, *8*, 86693. [CrossRef] [PubMed]
128. An, X.; Yu, W.; Liu, J.; Tang, D.; Yang, L.; Chen, X. Oxidative cell death in cancer: Mechanisms and therapeutic opportunities. *Cell Death Dis.* **2024**, *15*, 556. [CrossRef] [PubMed]
129. Rangsinth, P.; Sharika, R.; Pattarachotanant, N.; Duangjan, C.; Wongwan, C.; Sillapachaiyaporn, C.; Nilkhet, S.; Wongsirojkul, N.; Prasansuklab, A.; Tencomnao, T.; et al. Potential beneficial effects and pharmacological properties of ergosterol, a common bioactive compound in edible mushrooms. *Foods* **2023**, *12*, 2529. [CrossRef] [PubMed]
130. Sutthisa, W.; Anujakkawan, S. Antibacterial Potential of Oyster Mushroom (*Pleurotus ostreatus* (Jacq. Ex Fr.) P. Kumm.) Extract against Pathogenic Bacteria. *J. Pure Appl. Microbiol.* **2023**, *17*, 1907–1915. [CrossRef]
131. Effiong, M.E.; Umeokwochi, C.P.; Afolabi, I.S.; Chinedu, S.N. Comparative antioxidant activity and phytochemical content of five extracts of *Pleurotus ostreatus* (oyster mushroom). *Sci. Rep.* **2024**, *14*, 3794. [CrossRef] [PubMed]

132. Lesa, K.N.; Khandaker, M.U.; Mohammad Rashed Iqbal, F.; Sharma, R.; Islam, F.; Mitra, S.; Emran, T. Nutritional Value, Medicinal Importance, and Health-Promoting Effects of Dietary Mushroom (*Pleurotus ostreatus*). *J. Food Qual.* **2022**, *2022*, 2454180. [CrossRef]
133. Zhang, S.; Lin, L.; Yun, Z.; Ye, F.Y.; Zhao, G. Roles of mushroom polysaccharides in chronic disease management. *J. Integr. Agric.* **2022**, *21*, 1839–1866. [CrossRef]
134. Das, A.; Chen, C.M.; Mu, S.C.; Yang, S.H.; Ju, Y.M.; Li, S.C. Medicinal Components in Edible Mushrooms on Diabetes Mellitus Treatment. *Pharmaceutics* **2022**, *14*, 436. [CrossRef]
135. Erdoğan Eliuz, E.A. Antibacterial activity and antibacterial mechanism of ethanol extracts of *Lentinula edodes* (Shiitake) and *Agaricus bisporus* (button mushroom). *Int. J. Environ. Health Res.* **2022**, *32*, 1828–1841. [CrossRef] [PubMed]
136. Törős, G.; El-Ramady, H.; Prokisch, J.; Velasco, F.; Llanaj, X.; Nguyen, D.H.H.; Peles, F. Modulation of the Gut Microbiota with Prebiotics and Antimicrobial Agents from *Pleurotus ostreatus* Mushroom. *Foods* **2023**, *12*, 2010. [CrossRef] [PubMed]
137. Ruilova, M.; Niño-Ruiz, Z.; Sanbria, J.; Montero, D.; Salazar, S.; Bayas, F.; Sandoval, R. Antimicrobial activity of Lentinula edodes mushroom extracts against pathogenic bacteria. *Ital. J. Food Sci.* **2019**, *31*, 179–189.
138. Baptista, F.; Campos, J.; Costa-Silva, V.; Pinto, A.R.; Saavedra, M.J.; Ferreira, L.M.; Rodrigues, M.; Barros, A.N. Nutraceutical potential of Lentinula edodes' spent mushroom substrate: A comprehensive study on phenolic composition, antioxidant activity, and antibacterial effects. *J. Fungi* **2023**, *9*, 1200. [CrossRef] [PubMed]
139. Nagulwar, M.M.; More, D.R.; Mandhare, L.L. Nutritional properties and value addition of mushroom: A review. *Pharma Innov.* **2020**, *9*, 395–398.
140. Farooq, M.; Rakha, A.; Hassan, J.U.; Solangi, I.A.; Shakoor, A.; Bakhtiar, M.; Khan, M.N.; Khan, S.; Ahmad, I.; Ahmed, S.; et al. Physicochemical and Nutritional Characterization of Mushroom Powder Enriched Muffins. *J. Innov. Sci.* **2021**, *7*, 110–120. [CrossRef]
141. Kaur, K.; Sharma, R.; Srivastava, I.; Kaur, S.; Mehrotra, R. Edible Mushrooms: Nature's superfood for health and wellbeing. *Int. J. Innov. Multidiscip. Res.* **2022**, *1*, 40–53.
142. Chen, C.; Han, Y.; Li, S.; Wang, R.; Tao, C. Nutritional, antioxidant, and quality characteristics of novel cookies enriched with mushroom (*Cordyceps militaris*) flour. *CYTA J. Food* **2021**, *19*, 137–145. [CrossRef]
143. Owheruo, J.O.; Edo, G.I.; Oluwajuyitan, D.T.; Faturoti, A.O.; Martins, I.E.; Akpoghelie, P.O.; Agbo, J.J. Quality evaluation of value-added nutritious biscuit with high antidiabetic properties from blends of wheat flour and oyster mushroom. *Food Chem. Adv.* **2023**, *3*, 100375. [CrossRef]
144. Spim, S.R.V.; Castanho, N.R.C.M.; Pistila, A.M.H.; Jozala, A.F.; Oliveira Júnior, J.M.; Grotto, D. Lentinula edodes mushroom as an ingredient to enhance the nutritional and functional properties of cereal bars. *J. Food Sci. Technol.* **2021**, *58*, 1349–1357. [CrossRef]

foods

MDPI

Article

(1→3)-α-D-Glucan from the Pink Oyster Mushroom (*Pleurotus djamor*): Structural Features

Paulina Adamczyk [1], Iwona Komaniecka [2], Marek Siwulski [3], Kamila Wlizło [1], Adam Junka [4], Artur Nowak [1], Dariusz Kowalczyk [5], Adam Waśko [6,*], Jolanta Lisiecka [3], Michał Grzymajło [7] and Adrian Wiater [1,*]

1 Department of Industrial and Environmental Microbiology, Institute of Biological Sciences, Faculty of Biology and Biotechnology, Maria Curie-Skłodowska University, Akademicka 19, 20-033 Lublin, Poland; paulina.adamczyk@mail.umcs.pl (P.A.); kamila.wlizlo@mail.umcs.pl (K.W.); artur.nowak@mail.umcs.pl (A.N.)
2 Department of Genetics and Microbiology, Institute of Biological Sciences, Faculty of Biology and Biotechnology, Maria Curie-Skłodowska University, Akademicka 19, 20-033 Lublin, Poland; iwona.komaniecka@mail.umcs.pl
3 Department of Vegetable Crops, Faculty of Agriculture, Horticulture and Biotechnology, Poznań University of Life Sciences, Dąbrowskiego 159, 60-594 Poznań, Poland; marek.siwulski@up.poznan.pl (M.S.); jolanta.lisiecka@up.poznan.pl (J.L.)
4 "P.U.M.A.", Platform for Unique Model Application, Department of Pharmacy, Wroclaw Medical University, Borowska 211, 50-534 Wroclaw, Poland; adam.junka@umw.edu.pl
5 Department of Biochemistry and Food Chemistry, Faculty of Food Sciences and Biotechnology, University of Life Sciences in Lublin, Skromna 8, 20-704 Lublin, Poland; dariusz.kowalczyk@up.lublin.pl
6 Department of Biotechnology, Microbiology and Human Nutrition, Faculty of Food Science and Biotechnology, University of Life Sciences in Lublin, Skromna 8, 20-704 Lublin, Poland
7 Department of Polymer Engineering and Technology, Faculty of Chemistry, Wrocław University of Science and Technology (WUST); Wyb. Wyspiańskiego 27, 50-370 Wrocław, Poland; michal.grzymajlo@pwr.edu.pl
* Correspondence: adam.wasko@up.lublin.pl (A.W.); adrian.wiater@mail.umcs.pl (A.W.)

Abstract: (1→3)-α-D-Glucan is an important component of the cell wall of most fungi. The polymer has many applications, including as a therapeutic agent in the prevention or treatment of various diseases, as well as a heavy metal sorbent and a component of new materials used in the plastics industry. The presence of (1→3)-α-D-glucan (water-insoluble, alkali-soluble polysaccharide) in the cell wall of *Pleurotus djamor* (pink oyster mushroom) was confirmed using specific fluorophore-labeled antibodies. Therefore, the water-insoluble fraction (WI-ASF) of *P. djamor* B123 fruiting bodies was isolated by alkaline extraction and used for further analyses. The structural features of the WI-ASF were determined by composition analysis, linkage analysis, Fourier transform infrared and Raman spectroscopy, ^{1}H and ^{13}C nuclear magnetic resonance spectroscopy, scanning electron microscopy, as well as viscosity, specific rotation, and gel permeation chromatography. These studies revealed the presence of glucose units linked by α-glycosidic bonds and scanty amounts of mannose and xylose. Furthermore, methylation analysis of WI-ASF demonstrated that the (1→3)-linked glucopyranose (Glc*p*) is the primary moiety (86.4%) of the polymer, while the 3,4- and 3,6-substituted hexoses are the branching residues of the glucan. The results of chemical and spectroscopic investigations indicated that the analyzed WI-ASF is a (1→3)-linked α-D-glucan type with a molecular weight of 552 kDa.

Keywords: *Pleurotus djamor*; pink oyster mushroom; fruiting bodies; (1→3)-α-D-glucan; structural analyses

1. Introduction

(1→3)-α-D-Glucans are an essential component of the cell wall of most fungi (exceptions including the yeasts *Saccharomyces cerevisiae* and *Candida albicans*), where they have many biological functions, e.g., structural, nutritive, aggregation-promoting, or virulence factors for some pathogens (e.g., α-glucan from *Aspergillus fumigatus*) [1–3]. In comparison to other polymers present in the fungal cell wall, the (1→3)-α-D-glucans are relatively poorly understood and researched. The insoluble nature, the unique crystalline structure, and the presence of α-(1→3)-glycosidic linkages have defined the applicability of this polymer. To date, it has been shown, among other things, that (1→3)-α-D-glucans: (i) constitute a safe, inexpensive and readily available mutanase inducer, enzymes used in dental plaque degradation and prevention of tooth decay in humans and animals; (ii) after chemical modification (e.g., carboxymethylation, sulfation, aminopropylation, or hydroxyethylation) as water-soluble derivatives, they had significant anticancer, immunomodulatory, and antioxidant potential; (iii) can be used in combination with a catalyst (such as lipase) in the esterification reaction of higher fatty acids with alcohols with hydrophobic alkyl chains; (iv) can be an efficient sorbents for the removal of toxic metal ions from aquatic environments; (v) can be used to bind mycotoxins present in animal feed; (vi) are a source of oligosaccharides with prebiotic properties; and (vii) due to their exceptional thermoplastic and texture-forming properties, they are used in the plastics and textile industries [2,4–9].

The fruiting bodies of cultivated mushrooms, especially *Pleurotus* spp., are excellent sources of (1→3)-α-D-glucans. As shown in our previous studies, the biomass of these fungi was a source of glucans with anticancer properties [10,11]. An additional benefit is that waste fungal biomass, which is a crucial problem in the cultivation of oyster mushrooms, can be used for this purpose. This approach is in line with the zero-waste concept.

Pleurotus djamor (Rumph. ex Fr.) Boedijn is a species from the oyster mushroom group native to South-East Asia and Central America [12]. Salomones et al. [13] reported that the species also occurs in Africa and Australia. In tropical and subtropical regions, it occurs naturally on various species of trees, such as palms, bamboo, rubber trees, and mangoes, and is therefore readily available on the market [12,14,15]. Its cultivation is becoming increasingly popular worldwide due to the pink color of its fruiting bodies (pink oyster mushroom) and its high nutritional and medicinal value. Several studies have confirmed the antioxidant activity of *P. djamor* [16–19] as well as its broad spectrum of other medicinal activity, including hypoglycemic, anti-cancer, and anti-inflammatory effects [20–22]. Antimicrobial activity has also been demonstrated against different taxons of bacteria, e.g., *Escherichia coli*, *Staphylococcus aureus*, *Vibrio cholera*, *Xanthomonas* spp., and *Pseudomonas* spp. [23–25]. A positive effect of *P. djamor* on the human gut microbiota was also confirmed [26].

P. djamor is quite easy to cultivate due to its fast-growing mycelium and the possibility of using a wide range of waste materials derived from agriculture, forestry, and the food industry [27–32]. The fruiting bodies of this species are distinguished by their high dry matter content and very attractive aroma [33,34]. The high protein and the macro and micronutrient content of *P. djamor* accounts for its nutritional value [27,35]. The content of nutritionally important minerals in the fruiting bodies can easily be increased by supplementing the growing medium with some microelements, such as magnesium, zinc, and selenium salts [36–38].

As revealed by an analysis of the available articles, the amount of research work on *Pleurotus* (1→3)-α-D-glucans is limited. In this paper, we focus on the detection, isolation, structure, and some physicochemical properties of a water-insoluble (1→3)-α-D-glucan obtained from the pink oyster mushroom *P. djamor* via alkaline extraction.

2. Materials and Methods

2.1. Fungal Material and Fruiting Conditions

Strain B123 of *Pleurotus djamor* originated from the Mushroom Collection of the Department of Vegetable Crops at the University of Life Sciences in Poznań, Poland.

Cultivation of *P. djamor* fruiting bodies was carried out in the vegetation hall of the Department of Vegetable Crops at the University of Life Sciences in Poznań. The substrate for cultivation was prepared from a mixture of beech and alder sawdust (1:1, v/v) with several additives. One kilogram of the substrate contained 300 g of sawdust, 78 g of wheat bran, 20 g of maize flour, 2 g of gypsum, and 600 mL of tap water. The cultivation was carried out in polypropylene bottles (850 mL volume). Each bottle was filled with 500 g of the substrate. The bottles were sterilized at a temperature of 121 °C for 1 h. Mycelium of *P. djamor* on wheat grain was used for substrate inoculation in the amount of 10 g per bottle. Incubation was conducted at 25 °C and 80–85% relative air humidity until the substrate became completely covered with mycelium. Next, the bottles were placed in the cultivation chamber. For fructification, air relative humidity and temperature were maintained at 85–90% and 22 ± 1 °C, respectively. The cultivation was additionally lighted with fluorescent light of 500 lx intensity with a photoperiod regime of 12 h dark and 12 h light. The cultivation chamber was aerated in such a way as to maintain the CO_2 concentration below 1000 ppm. Fruiting bodies of *P. djamor* were harvested successively as they matured (Figure 1). The yield included whole fruiting bodies. The obtained fruiting bodies were dried in an electric drier at 40 °C to a constant weight.

Figure 1. Fruiting bodies of *P. djamor* B123 on an artificial substrate.

2.2. WI-ASF Isolation

The water-insoluble, alkali-soluble fraction (WI-ASF) was isolated from the fruiting bodies of *P. djamor* B123 according to the method described previously [39]. Briefly, the dried fruiting bodies (100 g) were milled into powder using a high-speed blade mill WŻ-1 (Research Institute for the Bakery Industry, Bydgoszcz, Poland) and extracted successively with 2 L of methanol, NaCl (0.9%), hot water, Na_2CO_3 (5%), and finally, with 1 M NaOH containing 0.02% sodium borohydride ($NaBH_4$, a reducing agent) for 24 h at room temperature (all reagents were form POCh, Gliwice, Poland). The final extract was neutralized with 1 M HCl under constant mixing, and the precipitated fraction (WI-ASF) was repeatedly washed with water, collected by centrifugation, and lyophilized. The yield of WI-ASF isolated from fruiting bodies of *P. djamor* B123 was 2.8 g/100 g of dry mass.

2.3. Sugar Composition and Methylation Analysis of WI-ASF from *P. djamor*

The WI-ASF (2 mg) was acid hydrolyzed using 2 M trifluoroacetic acid (TFA, Sigma-Aldrich, Saint Luis, MO, USA) at 100 °C for 4 h. Liberated monosaccharides were reduced by using sodium borodeuteride ($NaBD_4$, Sigma-Aldrich, Saint Luis, MO, USA), and obtained products were acetylated with pyridine-acetic anhydride mixture (Sigma-Aldrich, Saint Luis, MO, USA) (1:1, *v:v*) at 85 °C for 30 min to obtain alditol acetates [40]. For quantitative analyses, inositol was used as an internal standard.

Linkage analysis of sugars in the polysaccharide was performed with the method described by Hakomori [41]. Briefly, the sample (5 mg) was dissolved in 0.5 mL dry dimethylsulfoxide (DMSO, Sigma-Aldrich, Saint Luis, MO, USA), then 0.5 mL of dimsyl Na (de-protonating agent, made "on-side" from DMSO and NaH) was added, and methylation occurred in a presence of methyl iodide (1 mL, Sigma-Aldrich, Saint Luis, MO, USA). An excess of reagents was removed in a stream of nitrogen, and the reaction was stopped by adding water. The partly methylated polymer obtained was extracted into chloroform, dried in a stream of nitrogen, and acid hydrolyzed (2 M TFA, 100 °C for 4 h). The partly methylated sugar monomers were reduced and acetylated as above.

The absolute configuration of the monosaccharides was established by analysis of acetylated R-(−)2-butylglycosides according to the procedure described by Gerwig and co-workers [42].

All sugar derivatives were analyzed by GC-MS using a 7890A gas chromatograph connected to an MSD 5975C (inert XL EI/CI MSD) detector (Agilent Technologies, Santa Clara, CA, USA) equipped with an HP-5MS column (30 m × 0.25 mm, film thickness 0.25 µm; Agilent J&W GC Columns, Agilent, Santa Clara, CA, USA). Helium was a carrier gas with a flow rate of 1 mL/min. The temperature program was as follows: from 150 °C (3 min) raised to 320 °C at 5 °C min^{-1}, and the final temperature was maintained for 10 min. Sugar derivatives were identified based on their retention times and characteristic mass spectra, using Enhanced Data Analysis software, version 5.02.09 (Agilent Technologies, Santa Clara, CA, USA).

2.4. Miscellaneous Methods

Immunofluorescent labeling of the (1→3)-α-D-glucan in the cell wall of *P. djamor* B123 was carried out using specific antibodies, i.e., mouse IgM MOPC-104E and goat anti-mouse IgM labeled with a fluorescent dye [43]. Briefly, fresh mycelium of *P. djamor* B123 on Lab-Tek II Chamber slides (Nunc, Rochester, NY, USA) was fixed with a 3% (*v*/*v*) formaldehyde solution in distilled water at 65 °C for 30 min. The fixed fungal cells were washed three times in PBS (Phosphate Buffered Saline, pH 7.4) before being infiltrated with 1% (*v*/*v*) Tween 20 in PBS (PBS-T). To detect the presence of (1→3)-α-D-glucan, a 150 µL solution of mouse IgM MOPC-104E (0.1 mg/mL in PBS) (Sigma, Saint Louis, MO, USA) as a primary antibody and 150 µL Alexa Fluor 488 goat anti-mouse IgM (µ-chain specific) (0.1 mg/mL in PBS) (Sigma, Saint Louis, MO, USA) as a secondary antibody were used. Fungal material was incubated with the primary antibodies overnight at 4 °C in a wet chamber. Incubation with the secondary antibodies was performed for 2 h in the dark at 37 °C. Before observation the antibody-labeled cells were rinsed three times with PBS. The (1→3)-D-α-glucan was observed using an Olympus BX 51 microscope (Olympus, Tokyo, Japan) (excitation at 470/500 nm and emission at 525/550 nm).

The polysaccharide average molecular weight (Mw) was determined by gel permeation chromatography (GPC) using a Sepharose CL-6B column (Sigma-Aldrich, Saint Luis, MO, USA) (0.7 cm × 90 cm), as previously described by Choma et al. [44]. The (1→3)-α-D-glucan (2 mg) was dissolved in 0.5 mL of 1 M sodium hydroxide solution (POCh, Gliwice, Poland). The column was eluted with 1 M NaOH at a flow rate of 0.3 mL/min.

The separation was completed at room temperature. Due to the strongly alkaline character of eluate, the total content of carbohydrates was determined using the phenol-sulfuric acid assay according to the Dubois method [45]. The column was calibrated using dextrans of known molecular masses.

The Fourier transform infrared (FTIR) spectrum between 400 and 4000 cm^{-1} was recorded using a PerkinElmer FTIR spectrophotometer Model 1725X (PerkinElmer Corp., Norwalk, CT, USA). A specimen was prepared using the KBr-disk method.

The Fourier transform Raman (FT-Raman) spectrum was recorded using a Nicolet NXR FT-Raman module for a Nicolet 6700 Fourier transform infrared spectroscopy bench that used a InGaAs detector and CaF_2 beam splitter (Thermo Scientific, Madison, WI, USA). The sample in the form of powder was placed in stainless cubes and was illuminated using an Nd:YAG excitation laser operating at 1064 nm. The maximum laser power was 0.6 W. The spectrum was recorded in the range of 3700–150 cm^{-1}, and each spectrum averaged 200 scans with 8 cm^{-1} resolution. The analyzed spectra were averaged over three registered spectra.

The specific rotation $[\alpha]_D^{25}$ (*c* 1 M sodium hydroxide) was measured at 589 nm in a Perkin-Elmer Automatic Polarimeter model 341 LC (Wellesley, MA, USA). The viscosity of polysaccharides (*c* 1 M sodium hydroxide) was measured using a Brookfield model DV 3 viscometer (Stoughton, MA, USA) at 20 °C.

One-dimensional ^1H and ^{13}C NMR spectra were recorded with a Bruker Avance (300 MHz, Billerica, MA, USA) spectrometer at 60 °C (333 K) using standard Bruker software (TopSpin 3.2). The WI-ASF sample (5 mg) was dissolved in 0.7 mL of 1 M NaOD in D_2O (Sigma-Aldrich, Saint Louis, MO, USA). The ^1H and ^{13}C resonances were shifted to internal acetone (δ_H 2.225 ppm/δ_C 31.07 ppm).

Scanning electron microscopy (SEM) was conducted using an EVO Zeiss MA microscope. To prepare the (1→3)-α-D-glucan powder sample for SEM, a small amount of glucan lyophilizate was applied to a conductive double-sided adhesive tape affixed to a standard SEM stub. Excess unbound material was gently removed using a stream of compressed air. To ensure electrical conductivity and minimize charging effects, the sample was coated with a thin layer of an Au:Pd mixture using a Vacuum Sputter Coater. Observations were performed at an accelerating voltage of 10 kV with a working distance of 10 mm. Images were acquired at magnifications of 100×, 600×, and 2020× facilitating detailed assessment of the morphology and surface structure of the (1→3)-α-D-glucan samples. All measurements were carried out in high-vacuum conditions to ensure optimal image quality.

3. Results and Discussion

The presence of (1→3)-α-D-glucan in the cell wall of *P. djamor* B123 was confirmed using specific antibodies, i.e., primary mouse IgM MOPC-104E and goat anti-mouse IgM labeled with fluorescent dye AlexaFluor 488. Figure 2 shows the results of this investigation. The prominent green luminescence seen in panel **b** indicates the existence of (1→3)-α-D-glucan in the cell wall structure of the *P. djamor* filaments. As can be seen, the (1→3)-α-D-glucan is found along the entire length of the filament, suggesting that this polymer plays an important role in the organization of the fungal cell wall. Numerous studies indicate that (1→3)-α-D-glucan is differentially localized in the fungal cell wall. Studies carried out by Grün in 2003 [46] showed that this polymer is located under the layer of β-glucan, just above the cytoplasmic membrane. However, as suggested by recent reports on the fungal cell wall structure, the position of (1→3)-α-D-glucans is quite variable and species-dependent [1,47]. They can be found in the outer layers of the cell wall (in *Histoplasma capsulatum*) or as part of a complex polysaccharide network under the protein-polysaccharide layer (in *Aspergillus fumigatus*). Furthermore, the distribution of (1→3)-α-D-glucans within the cell wall is influenced by the developmental form of the fungus, i.e., spores or vegetative mycelium, and the type of microorganism culture [47,48].

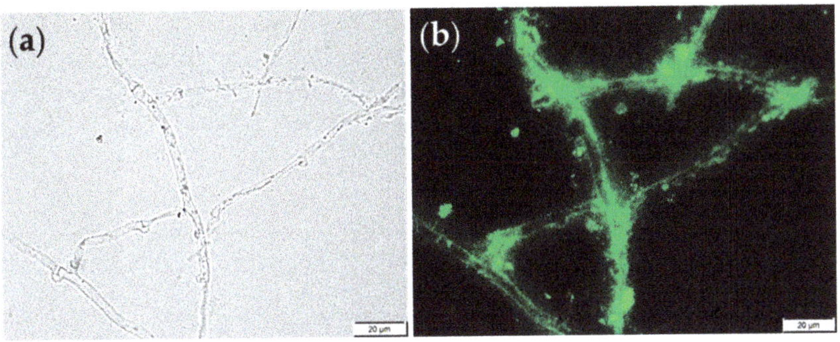

Figure 2. Immunofluorescent labeling of (1→3)-α-D-glucan in the mycelium of *P. djamor* B123 using specific labeled antibodies. (**a**) View in the light microscopy; (**b**) The same filaments in fluorescence microscopy. At least 10 preparations were observed, and representative images were chosen for presentation. Scale bar corresponds to 20 μm.

The WI-ASF used for further analysis of structural characteristics was isolated via alkaline extraction from the fruiting bodies of *P. djamor* B123 grown on artificial medium (Figure 1). The alkali-soluble polysaccharide material (WI-ASF) was obtained with a yield of 2.8% of dry mass. The WI-ASF contained mainly carbohydrates (97.5% of isolated material), as estimated by Dubois method [45], and only traces of protein (0.33%) were detected, as determined using the Bradford method [49].

The GC-MS analysis of monosaccharides liberated by the acid hydrolysis of the WI-ASF isolated from the fruiting bodies of *P. djamor* showed that the isolated material was composed predominantly of glucose (93.5%), suggesting that this is a glucan polymer. Only small amounts of the other sugars were detected, namely xylose (4.0%) and mannose (2.5%) (Figure 3). All the monosaccharides released have been shown to have the D absolute configuration [42].

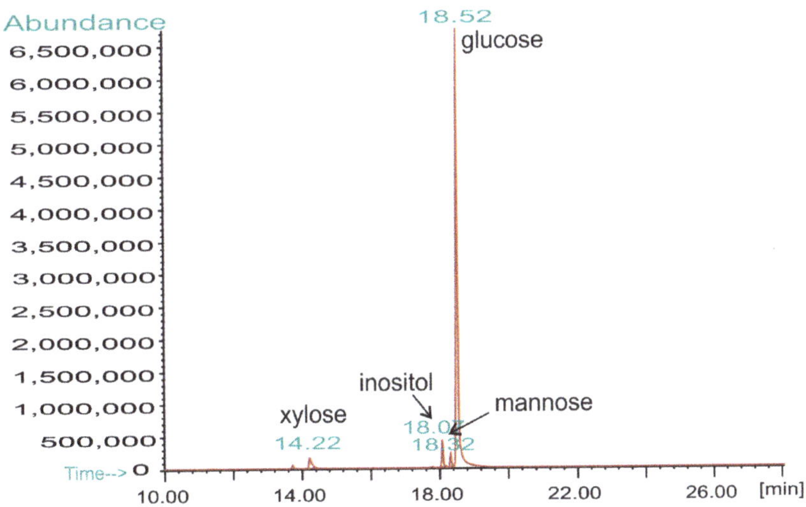

Figure 3. GC-MS chromatogram of monosaccharides released from *P. djamor* WI-ASF. Inositol was used as an internal standard.

The methylation analysis of the WI-ASF showed that (1→3)-linked Hex*p* (identified as (1→3)-linked glucose) is the major chain constituent (86.4%), while (1→4)-linked Hex*p* (4.5%) is the minor one (Table 1). It can be assumed that the isolated polysaccharide is (1→3)-glucan, with a rather short side chains composed of 4-linked hexose. The analyzed WI-ASF also contained two types of doubly substituted glucose residues, i.e., 3,4)-Hex*p* and 3,6)-Hex*p*, representing the main branching points of the (1→3)-linked polysaccharide chain. Moreover, fully methylated (terminal) xylose (2.7%) and hexose (2.1%) as well as (1→2)-linked pentose derivatives were also present.

Table 1. Results of the methylation analysis of the *P. djamor* WI-ASF polysaccharide.

Methylated Sugar	Linkage Type	Mol (%)
2,3,4-*O*-Me_3-pentose	*t*Pen*p*-(1→	2.7
3,4-*O*-Me_2-pentose	→2)-Pen*p*-(1→	1.3
2,3,4,6-*O*-Me_4-hexose	*t*Hex*p*-(1→	2.1
2,4,6-*O*-Me_3-hexose	→3)-Hex*p*-(1→	86.4
2,3,6-*O*-Me_3-hexose	→4)-Hex*p*-(1→	4.5
2,6-*O*-Me_2- hexose	→3,4)-Hex*p*-(1→	1.2
2,4-*O*-Me_2- hexose	→3,6)-Hex*p*-(1→	1.8

The polysaccharide preparation was methylated before being hydrolyzed, reduced, and acetylated. Therefore, the methylated sugars mentioned refer to permethylated alditol acetates identified by GC–MS (e.g., 2,3,4,6-*O*-Me_4-hexose refers to 1,5-di-*O*-acetyl-2,3,4,6-tetra-*O*-methyl-hexitol). *t*–terminal residue.

In summary, the composition analyses indicated that the monosaccharide residues (almost exclusively of α-D-glucose) in the WI-ASF isolated from the fruiting bodies of *P. djamor* are mainly (1→3)-linked.

The FTIR spectrum of the WI-ASF extracted from the *P. djamor* fruiting bodies is shown in Figure 4a. The wavelength of the FTIR spectrum was set in the range of 400–4000 cm^{-1}. The strong and broad band at 3334 cm^{-1} can be attributed to O–H stretching vibrations, and the weaker absorption band at 2931 cm^{-1} can be assigned to C–H vibrations. Furthermore, the peaks in the region of 950–1200 cm^{-1} attributed to the stretching vibrations of the pyranose ring and C-O-C and C-O-H moieties are characteristic for polysaccharides [50,51]. The WI-ASF showed absorption peaks at 931.50, 847.54, and 822.49 cm^{-1}. These peaks are characteristic of (1→3)-α-D-glucans [52]. In particular, the symmetrical band at 822.49 cm^{-1} is exclusively associated with (1→3)-α-linkages [53]. The low intense band visible at about 1640 cm^{-1} (derived from Amide I) may be indicative for the presence of very small amounts of proteins in the sample. The (1→3)-α-structure of the WI-ASF obtained from *P. djamor* was also confirmed by Raman spectroscopy (Figure 4b). Bands characteristic of polysaccharides are seen at 1460.81, 1360.00, 1255.30, 1125.43, and 1076.00 cm^{-1}. In turn, the intense and sharp bands at 941.04 and 549.77 cm^{-1} confirmed (1→3)-α-D-glucan as the major component of the WI-ASF [54,55].

To confirm the results of the compositional and spectral analyses of the WI-ASF and establish the anomeric configuration of the *P. djamor* polysaccharide, ^1H and ^{13}C NMR spectra were recorded in 1 M NaOD in D_2O. The 1D NMR spectra are shown in Figure 5. The only spin system identified was characteristic of α-D-glucose. The low-field anomeric proton signal (H-1) at δ 5.62 ppm and the anomeric carbon signal (C-1) at δ 101.2 ppm are typical of the α-anomer of (1→3)-linked glucose [56]. Furthermore, the intense and sharp peak near 71.7 ppm and smaller ones at 83.8 and 62.4 ppm correspond, respectively, to C-2, C-3 and C-6 carbons of (1→3)-α-D-glucan [55].

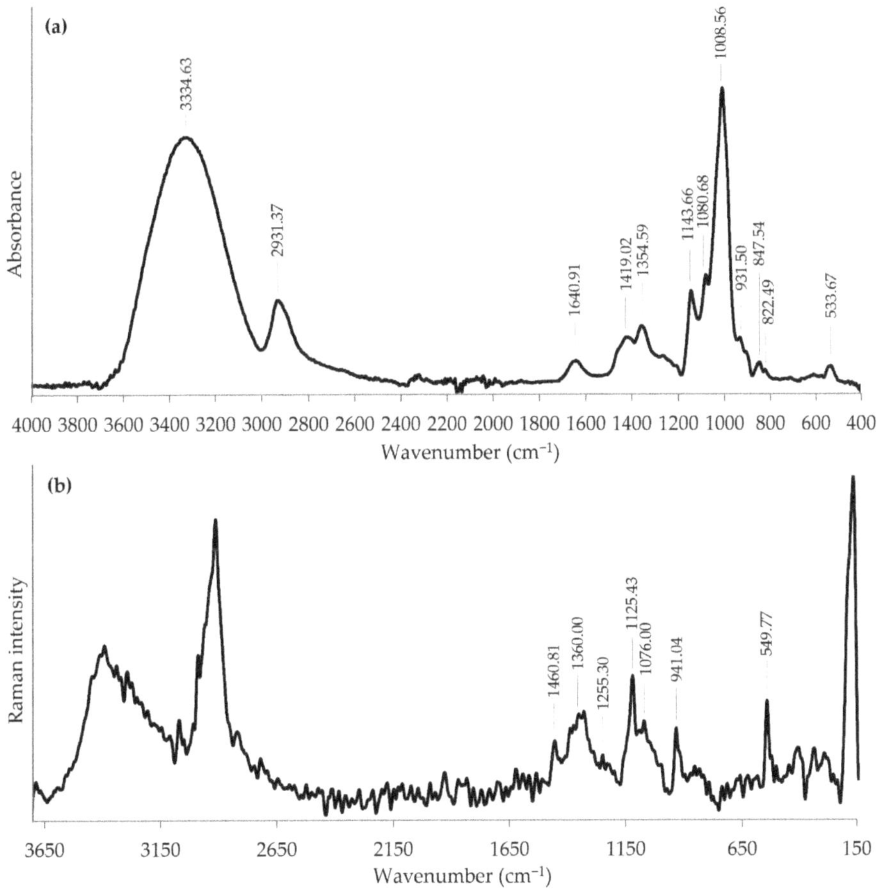

Figure 4. FTIR (**a**) and FT-Raman (**b**) spectra of WI-ASF extracted from the fruiting bodies of *P. djamor*.

Gel permeation chromatography (GPC) allowed the average molecular weight (Mw) of the (1→3)-α-D-glucan to be determined (Figure 6). The polysaccharide preparation dissolved in 1.0 M NaOH and fractionated on a Sepharose CL-6B column exhibited a single broad peak, giving an average Mw of about 552 kDa in the maximum point. This result indicates that polysaccharide chains are long, and more than 3400 sugar subunits were counted. (1→3)-α-D-glucan with similar Mw (560 kDa) was also isolated from the fruiting bodies of *Agrocybe cylindracea* [57]. Molecular weight estimation of the alkali-soluble polysaccharide from *Boletus edulis* showed two peaks, a main at about 850 kDa and a smaller broad peak with a maximum at about 92 kDa [57]. High and low molecular weight fractions, i.e., 2900 and 110.3 kDa, and 2300 and 20.3 kDa, were also detected for alkaline extracts obtained from *Pleurotus ostreatus* and *P. eryngii* fruiting bodies, respectively [55].

The specific optical rotation ($[\alpha]_D^{25}$) of the WI-ASF was +220° (*c* 1.0, 1 M NaOH). The high and positive value of the optical rotation was comparable to (1→3)-α-D-glucans from *Aspergillus* spp. (from +216° to +384°), *Cerrena unicolor* (+206°), *Agrocybe cylindracea* (+195°), and *Lentinus edodes* (+193.5°) [2,53,58]. The WI-ASF viscosity value oscillated around 11.45 mPa·s, and was similar to values (from 12.2 to 17.0 mPa·s) measured for (1→3)-α-D-glucans isolated from *Aspergillus* spp. biomass, including the linear pseudonigeran (8.8 mPa·s) from *A. niger* [2].

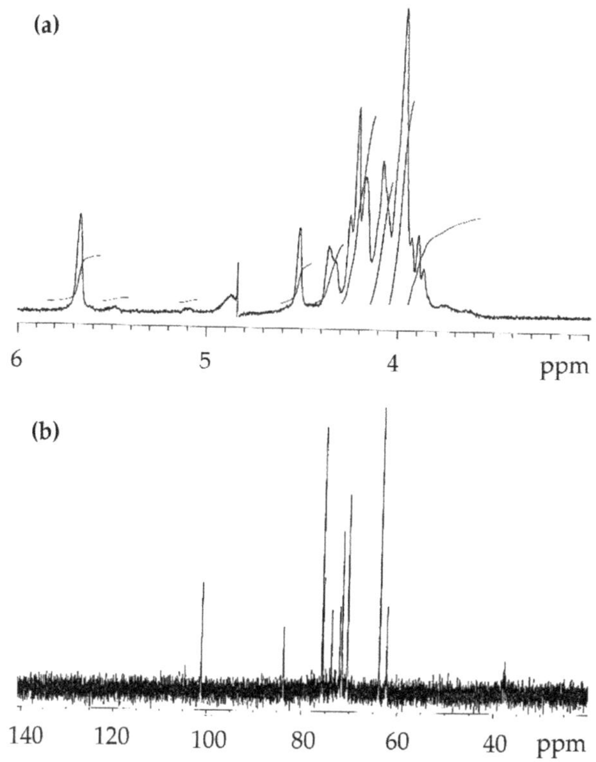

Figure 5. ^{1}H (**a**) and ^{13}C NMR (**b**) spectra of WI-ASF obtained from the fruiting bodies of *P. djamor*.

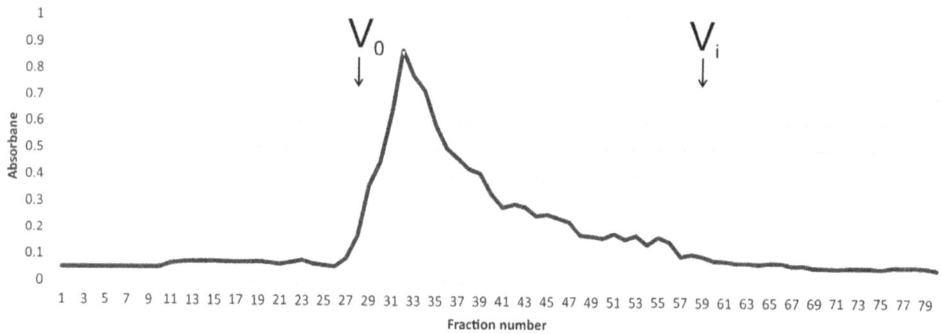

Figure 6. Elution profile (GPC) of the *P. djamor* WI-ASF at Sepharose CL-6B washed with 1.0 M NaOH. V_0—dead volume; V_i—inside volume.

Taking all the above data together, it can be concluded that the WI-ASF extracted from the fruiting bodies of *P. djamor* contains mainly (1→3)-α-D-glucan.

The scanning electron microscopy (SEM) analysis was used to complete the (1→3)-α-D-glucan characterization. The SEM images show the polysaccharide microstructure at magnifications of 100×, 600×, and 2020× (Figure 7). At 100× magnification, the image reveals the overall spatial arrangement of (1→3)-α-D-glucan, organized as a heterogeneous fibrous network (Figure 7a). This network results from intertwined polysaccharide fibers that constitute (1→3)-α-D-glucan. Increasing the magnification to 600× facilitates detailed observation of the individual fibers forming the (1→3)-α-D-glucan structure (Figure 7b).

The image shows interconnected irregular fibers creating a three-dimensional matrix with uneven pores. Notably, the fibers exhibit varying diameters, with some appearing thicker and others more thread-like. At the highest magnification of 2020×, the SEM image highlights micropores and the surface intricacies of the fibers (Figure 7c). The (1→3)-α-D-glucan exhibit porous architecture, which may influence its functional properties, such as water retention or interaction with other substances. The surface displays numerous indentations and pores, indicating a substantial specific surface area. In summary, these SEM images illustrate that the (1→3)-α-D-glucan extracted from *P. djamor* forms a complex fibrous network with irregular porosity.

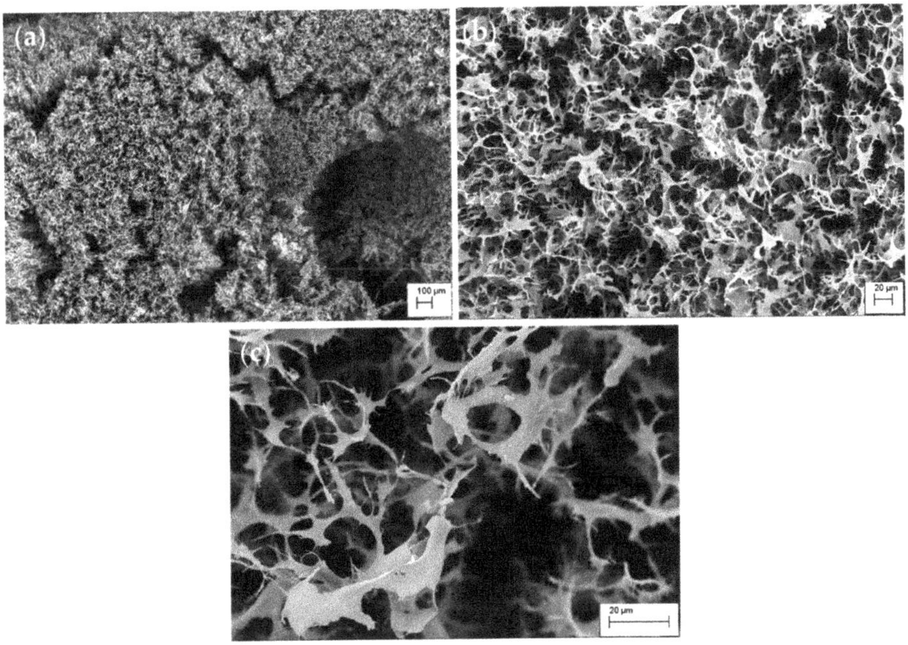

Figure 7. SEM of (1→3)-α-D-glucan from the fruiting bodies of *P. djamor*: (**a**) ×100, (**b**) ×600, (**c**) ×2020 magnification.

4. Conclusions

This study showed that (1→3)-α-D-glucan is the structural component of the cell wall of pink oyster mushroom *P. djamor*. All the analyses performed confirm that the water-insoluble, alkali-soluble polysaccharidic material extracted from the fruiting bodies of *P. djamor* B123 contained mainly (1→3)-α-D-glucan. A number of reports indicate that (1→3)-α-D-glucan can be used in a wide range of applications [2,59–61]; therefore, new cheap and safe sources of this polymer are being sought. The occurrence of (1→3)-α-D-glucan in *P. djamor* biomass opens up the possibility of extracting this polymer on a larger scale from cultivated oyster mushrooms, especially those considered a waste product.

Author Contributions: Conceptualization, P.A. and A.W. (Adrian Wiater); methodology, I.K., M.S., A.J., D.K. and A.W. (Adrian Wiater); formal analysis, P.A., I.K., M.S., K.W., A.J., A.N., D.K., A.W. (Adam Waśko), J.L., M.G. and A.W. (Adrian Wiater); investigation, P.A., I.K., M.S., K.W., A.J., A.N., D.K., A.W. (Adam Waśko), J.L., M.G. and A.W. (Adrian Wiater); resources, M.S. and J.L.; data curation, A.W. (Adrian Wiater); writing—original draft preparation, P.A., I.K., M.S., K.W., A.J., A.N., D.K., A.W. (Adam Waśko), J.L., M.G. and A.W. (Adrian Wiater); writing—review and editing, P.A., I.K. and A.W. (Adrian Wiater); visualization, P.A., I.K., M.S., A.J. and A.W. (Adrian Wiater); supervision, A.W.

(Adrian Wiater); project administration, P.A. and A.W. (Adrian Wiater); funding acquisition, A.W. (Adrian Wiater). All authors have read and agreed to the published version of the manuscript.

Funding: This research received no external funding.

Institutional Review Board Statement: Not applicable.

Informed Consent Statement: Not applicable.

Data Availability Statement: The original contributions presented in this study are included in the article. Further inquiries can be directed to the corresponding authors.

Conflicts of Interest: The authors declare no conflicts of interest.

Abbreviations

The following abbreviations are used in this manuscript:

WI-ASF	Water-insoluble alkali-soluble fraction
PBS	Phosphate buffered saline
SEM	Scanning electron microscopy
TFA	Trifluoroacetic acid
GPC	Gel permeation chromatography
FT-IR	Fourier transform infrared spectroscopy
FT-Raman	Fourier transform Raman

References

1. Ruiz-Herrera, J.; Ortiz-Castellanos, L. Cell wall glucans of fungi. A review. *Cell Surf.* **2019**, *5*, 100022. [CrossRef] [PubMed]
2. Złotko, K.; Wiater, A.; Waśko, A.; Pleszczyńska, M.; Paduch, R.; Jaroszuk-Ściseł, J.; Bieganowski, A. A report on fungal (1→3)-α-D-glucans: Properties, functions and application. *Molecules* **2019**, *24*, 3972. [CrossRef] [PubMed]
3. Shetty, P.R.; Batchu, U.R.; Buddana, S.K.; Rao, K.S.; Penna, S. A comprehensive review on α-D-Glucans: Structural and functional diversity, derivatization and bioapplications. *Carbohydr. Res.* **2021**, *503*, 108297. [CrossRef]
4. Wiater, A.; Pleszczyńska, M.; Szczodrak, J.; Janusz, G. Comparative studies on the induction of *Trichoderma harzianum* mutanase by α-(1→3)-glucan-rich fruiting bodies and mycelia of *Laetiporus sulphureus*. *Int. J. Mol. Sci.* **2012**, *13*, 9584–9598. [CrossRef]
5. Wiater, A.; Waśko, A.; Adamczyk, P.; Gustaw, K.; Pleszczyńska, M.; Wlizło, K.; Skowronek, M.; Tomczyk, M.; Szczodrak, J. Prebiotic potential of oligosaccharides obtained by acid hydrolysis of α-(1→3)-glucan from *Laetiporus sulphureus*: A pilot study. *Molecules* **2020**, *25*, 5542. [CrossRef]
6. Huang, Q.; Zhang, L. Preparation, chain conformation and anti-tumor activities of water-soluble phosphated (1→3)-α-D-glucan from *Poria cocos* mycelia. *Carbohydr. Polym.* **2011**, *83*, 1363–1369. [CrossRef]
7. Araujo, R.V.S.; Melo-Junior, M.R.; Beltrao, E.I.C.; Mello, L.A.; Iacomini, M.; Carneiro-Leao, A.M.A.; CarvalhoJr, L.B.; Satos-Magalhaes, N.S. Evaluation of the antischistosomal activity of sulfated α-D-glucan from the lichen *Ramalina celastri* free and encapsulate into liposomes. *Braz. J. Med. Biol. Res.* **2011**, *44*, 311–318. [CrossRef]
8. Nowak, K.; Wiater, A.; Choma, A.; Wiacek, D.; Bieganowski, A.; Siwulski, M.; Wasko, A. Fungal (1→3)-α-D-glucans as a new kind of biosorbent for heavy metals. *Int. J. Biol. Macromol.* **2019**, *137*, 960–965. [CrossRef] [PubMed]
9. Puanglek, S.; Kimura, S.; Enomoto-Rogers, Y.; Kabe, T.; Yoshida, M.; Wada, M.; Iwata, T. *In vitro* synthesis of linear α-1,3-glucan and chemical modification to ester derivatives exhibiting outstanding thermal properties. *Sci. Rep.* **2016**, *6*, 30479. [CrossRef]
10. Wiater, A.; Paduch, R.; Próchniak, K.; Pleszczynska, M.; Siwulski, M.; Bialas, W.; Szczodrak, J. Ocena aktywności biologicznej karboksymetylowanych pochodnych α-(1→3)-glukanów wyizolowanych z owocników uprawnych gatunków boczniaka (*Pleurotus*). *Zywn. Nauk. Technol. Ja.* **2015**, *98*, 193–206. [CrossRef]
11. Wiater, A.; Paduch, R.; Pleszczyńska, M.; Próchniak, K.; Choma, A.; Kandefer-Szerszeń, M.; Szczodrak, J. α-(1→ 3)-D-Glucans from fruiting bodies of selected macromycetes fungi and the biological activity of their carboxymethylated products. *Biotechnol. Lett.* **2011**, *33*, 787–795. [CrossRef] [PubMed]
12. Stamets, P. *Growing Gourmet and Medicinal Mushrooms*, 3rd ed.; Ten Speed Press: Berkeley, CA, USA, 2011; pp. 295–300.
13. Salmones, D.; Valdez, L.M.; Gatian-Hernande, R. Entecruzamiento y evaluacion de la produccion de las variedades de *Pleurotus djamor* (Fr.) Boedijn. *Rev. Mex. Mic.* **2004**, *18*, 21–26. [CrossRef]
14. Arbaayah, H.H.; Umi, K.Y. Antioxidant properties in the oyster mushrooms (*Pleurotus* spp.) and split gill (*Scizophyllum commune*) ethanolic extracts. *Mycosphere* **2013**, *4*, 661–673. [CrossRef]

15. Srivastava, M. A pink coloured *Pleurotus djamor* (Rumph.) Boedijn from natural habitat of north Bihar, India. *Curr. Sci.* **2001**, *80*, 337–338.
16. İnci, Ş.; Akyüz, M.; Kirbağ, S. Antimicrobial, antioxidant, cytotoxicity and DNA protective properties of the pink oyster mushroom, *Pleurotus djamor* (Agaricomycetes). *Int. J. Med. Mushrooms* **2023**, *25*, 55–66. [CrossRef]
17. Vega, A.; De León, J.A.; Miranda, S.; Reyes, S.M. Agro-industrial waste improves the nutritional and antioxidant profile of *Pleurotus djamor*. *Clean. Waste Syst.* **2022**, *2*, 100018. [CrossRef]
18. Maity, G.N.; Maity, P.; Khatua, S.; Acharya, K.; Dalai, S.; Mondal, S. Structural features and antioxidant activity of a new galactoglucan from edible mushroom *Pleurotus djamor*. *Int. J. Biol. Macromol.* **2021**, *168*, 743–749. [CrossRef]
19. Raman, J.; Sivakumar, A.; Lakshmanan, H.; Raaman, N.; Shin, H.J. Antioxidant activity of partially characterized polysaccharides from the edible mushroom *Pleurotus djamor* var. roseus. *J. Mushroom.* **2021**, *19*, 140–149. [CrossRef]
20. Shreya, S.; Jain, S.K.; Guru, S.K.; Sahu, A.N. Anti-cancer potential of *Pleurotus* mushroom: Detailed insight on the potential bioactive molecules, *in vitro—in vivo* studies, and formulation. *Lett. Drug Des. Discov.* **2023**, *20*, 439–456. [CrossRef]
21. Li, H.; Feng, Y.; Sun, W.; Kong, Y.; Jia, L. Antioxidation, anti-inflammation and anti-fibrosis effect of phosphorylated polysaccharides from *Pleurotus djamor* mycelia on adenine-induced chronic renal failure mice. *Int. J. Biol. Macromol.* **2021**, *170*, 652–663. [CrossRef]
22. Nayak, H.; Kushwaha, A.; Behera, P.C.; Shahi, N.C.; Kushawaha, K.P.S.; Kumar, A.; Mishra, K.K. The pink oyster mushroom, *Pleurotus djamor* (Agaricomycetes): A potent antioxidant and hypoglycemic agent. *Int. J. Med. Mushrooms* **2021**, *23*, 29–36. [CrossRef] [PubMed]
23. Afsar, M.; Zia, A.; Salam, M.B.U.; Ahmad, M.N.; Khan, A.A.; Haq, T.; Aziz, T.; Alasmari, A.F. A multifaceted analysis of spent mushroom substrate of selected oyster mushrooms for enzymatic activity, proximate composition, and antimicrobial activity. *Ital. J. Food Sci.* **2024**, *36*, 165–174. [CrossRef]
24. Maikon, R.E.; Ávila, S.; Lima, J.J.; Silva, R.S.A.; Andrade, L.P.; Bacila, D.M.; Mathias, A.L.; Jorge, R.M.M. *Araucaria angustifolia* seed coat waste reduction through its utilization in substrate diversification for *Pleurotus djamor* production. *Sci. Hortic.* **2024**, *330*, 113060. [CrossRef]
25. Otali, C.C.; Otoikhian, C.S.O.; Onuoha, T.; Akpeji, C.S.; Bosah, B.O. Antibacterial activities of *Pleurotus ostreatus* and *Pleurotus djamor* against selected bacterial pathogens. *Bima J. Sci.Technol.* **2024**, *8*, 397–402. [CrossRef]
26. Andrade, G.M.; Souza, E.L.; Zárate-Salazar, J.R.; Oliveira, J.N.; Tavares, J.F.; Lima, M.S.; Medeiros, R.L.; Albuquerque, T.M.R.; Pereira, F.O. Unveiling the potential prebiotic effects of edible mushroom *Pleurotus djamor* during *in vitro* colonic fermentation. *J. Agric. Food Chem.* **2024**, *72*, 26722–26732. [CrossRef]
27. İnci, Ş.; Kirbağ, S.; Akyüz, M. Valorization of local agro-residues for the cultivation of Pleurotus djamor (Rumph. Ex Fr.) Boedijn and their effects on nutritional value. *Biomass Conv. Bioref.* **2024**, 1–10. [CrossRef]
28. Nguyen, B.T.T.; Le, V.V.; Nguyen, H.N.; Nguyen, H.T.T.; Nguyen, L.T.; Ngo, N.X. Cotton waste as an optimal substrate for cultivation of the pink oyster mushroom *Pleurotus djamor*. *J. App. Biol. Biotech.* **2025**, *13*, 184–191. [CrossRef]
29. Cruz-Moreno, B.A.; Pérez, A.A.F.; García-Trejo, J.F.; Pérez-García, S.A.; Gutiérrez-Antonio, C. Identification of secondary metabolites of interest in *Pleurotus djamor* using agave tequilana bagasse. *Molecules* **2023**, *28*, 557. [CrossRef]
30. Jegadeesh, R.; Lakshmanan, H.; Kab-Yeul, J.; Sabaratnam, V.; Raaman, N. Cultivation of pink oyster mushroom Pleurotus djamor var. roseus on various agro-residues by low cost technique. *J. Mycopathol. Res.* **2018**, *56*, 213–220.
31. Silva, L.A.; Dulay, R.M.R.; Kalaw, S.P. Mycelial growth of pink oyster mushroom (*Pleurotus djamor*) on banana sucrose gulaman and fruiting body production on banana-based substrate formulations. *CLSU Int. J. Sci. Technol.* **2018**, *3*, 24–32. [CrossRef]
32. Satpal, S.; Gopal, S.; Kumar, R. Effect of different substrates on the growth and yield of oyster mushroom (*Pleurotus djamor*). *Int. J. Agric. Sci.* **2017**, *9*, 3721–3723.
33. Yin, C.; Fan, X.; Fan, Z.; Shi, D.; Yao, F.; Gao, H. Comparison of non-volatile and volatile flavor compounds in six *Pleurotus* mushrooms. *J. Sci. Food Agric.* **2019**, *99*, 1691–1699. [CrossRef]
34. Zawirska-Wojtasiak, R.; Siwulski, M.; Mildner-Szkudlarz, S.; Wąsowicz, E. Studies on the aroma of different species and strains of *Pleurotus* measured by GC/MS, sensory analysis and electronic nose. *Acta Sci. Pol. Technol. Aliment.* **2009**, *8*, 47–61.
35. Raman, J.; Lakshmanan, H.; Jang, K.Y.; Oh, M.; Oh, Y.L.; Im, J.H. Nutritional composition and antioxidant activity of pink oyster mushrooms (*Pleurotus djamor* var. roseus) grown on a paddy straw substrate. *J. Mushroom* **2020**, *18*, 189–200. [CrossRef]
36. Madaan, K.; Sharma, S.; Kalia, A. Effect of selenium and zinc biofortification on the biochemical parameters of *Pleurotus* spp. under submerged and solid-state fermentation. *J. Trace Elem. Med. Biol.* **2024**, *82*, 127365. [CrossRef]
37. Sravani, A.; Sharma, S.; Kalia, A. Effect of selenium enriched wheat substrate on nutritional and antioxidant properties of *Pleurotus* spp. *Acta Aliment.* **2021**, *50*, 358–368. [CrossRef]
38. Zięba, P.; Sękara, A.; Bernaś, E.; Krakowska, A.; Sułkowska-Ziaja, K.; Kunicki, E.; Suchanek, M.; Muszyńska, B. Supplementation with magnesium salts—A strategy to increase nutarceutical value of *Pleurotus djamor* fruiting bodies. *Molecules* **2021**, *26*, 3273. [CrossRef]

39. Kiho, T.; Yoshida, I.; Katsuragawa, M.; Sakushima, M.; Usui, S.; Ukai, S. Polysaccharides in fungi XXXIV: A polysaccharide from fruiting bodies of *Amanita muscaria* and the antitumor activities of its carboxymethylated product. *Biol. Pharm. Bull.* **1994**, *17*, 1460–1462. [CrossRef]
40. Sawardeker, J.S.; Sloneker, J.H.; Jeanes, A.R. Quantitative determination of monosaccharides as their alditol acetates by gas liquid chromatography. *Anal. Chem.* **1965**, *37*, 1602–1604. [CrossRef]
41. Hakomori, S. Rapid permethylation of glycolipids and polysaccharides catalyzed by methyl carbon. *J. Biochem.* **1964**, *55*, 205–208.
42. Gerwig, G.J.; Kamerling, J.P.; Vliegenthart, J.F.G. Determination of the D and L configuration of neutral monosaccharides by high-resolution capillary g.l.c. *Carbohydr. Res.* **1978**, *62*, 349–357. [CrossRef]
43. Fujikawa, T.; Kuga, Y.; Yano, S.; Yoshimi, A.; Tachiki, T.; Abe, K.; Nishimura, M. Dynamics of cell wall components of *Magnaporthe grisea* during infectious structure development. *Mol. Microbiol.* **2009**, *73*, 553–570. [CrossRef]
44. Choma, A.; Wiater, A.; Komaniecka, I.; Paduch, R.; Pleszczyńska, M.; Szczodrak, J. Chemical characterization of a water insoluble (1→3)-α-D-glucan from an alkaline extract of *Aspergillus wentii*. *Carbohydr. Polym.* **2013**, *91*, 603–608. [CrossRef]
45. Dubois, M.; Gilles, K.A.; Hamilton, J.K.; Rebers, P.A.; Smith, F. Colorimetric method for determination of sugars and related substances. *Anal. Chem.* **1956**, *28*, 350–356. [CrossRef]
46. Grün, C.H. Structure and Bosynthesis of Fungal α-Glucans. Ph.D. Thesis, Universiteit Utrecht, Utrecht, The Netherlands, 2003.
47. Erwig, L.P.; Gow, N.A.R. Interactions of fungal pathogens with phagocytes. *Nat. Rev. Microbiol.* **2016**, *14*, 163–176. [CrossRef] [PubMed]
48. Beauvais, A.; Fontaine, T.; Aimanianda, V.; Latge, J.P. *Aspergillus* cell wall and biofilm. *Mycopathologia* **2014**, *178*, 371–377. [CrossRef]
49. Bradford, M. A rapid and sensitive method for the quantitation of microgram quantities of protein utilizing the principle of protein-dye binding. *Anal. Biochem.* **1976**, *72*, 248–254. [CrossRef]
50. Liu, C.; Li, X.; Li, Y.; Feng, Y.; Zhou, S.; Wang, F. Structural characterisation and antimutagenic activity of a novel polysaccharide isolated from *Sepiella maindroni* ink. *Food Chem.* **2008**, *110*, 807–813. [CrossRef]
51. Gieroba, B.; Kalisz, G.; Krysa, M.; Khalavka, M.; Przekora, A. Application of vibrational spectroscopic techniques in the study of the natural polysaccharides and their cross-linking process. *Int. J. Mol. Sci.* **2023**, *24*, 2630. [CrossRef]
52. Seymour, F.R.; Julian, R.L.; Jeanes, A.; Lamberts, B.L. Structural analysis of insoluble D-glucans by Fourier-transform, infrared difference-spectrometry: Correlation between structures of dextrans from strains of *Leuconostoc mesenteroides* and of D-glucans from strains of *Streptococcus mutans*. *Carbohydr. Res.* **1980**, *86*, 227–246. [CrossRef]
53. Zhang, P.; Zhang, L.; Cheng, S. Chemical structure and molecular weights of α-(1→3)-D-glucan from *Lentinus edodes*. *Biosci. Biotechnol. Biochem.* **1999**, *63*, 1197–1202. [CrossRef] [PubMed]
54. Edwards, H.G.M.; Russell, N.C.; Weinstein, R.; Wynn-Williams, D.D. Fourier transform Raman spectroscopic study of fungi. *J. Raman Spectrosc.* **1995**, *26*, 911–916. [CrossRef]
55. Synytsya, A.; Míčková, K.; Synytsya, A.; Jablonský, I.; Spěváček, J.; Erban, V.; Kováříková, E.; Čopíková, J. Glucans from fruit bodies of cultivated mushrooms *Pleurotus ostreatus* and *Pleurotus eryngii*: Structure and potential prebiotic activity. *Carbohydr. Polym.* **2009**, *76*, 548–556. [CrossRef]
56. Wang, T.; Deng, L.; Li, S.; Tan, T. Structural characterization of a water-insoluble (1→3)-α-D-glucan isolated from the *Penicillium chrysogenum*. *Carbohydr. Polym.* **2007**, *67*, 133–137. [CrossRef]
57. Choma, A.; Nowak, K.; Komaniecka, I.; Waśko, A.; Pleszczyńska, M.; Siwulski, M.; Wiater, A. Chemical characterization of alkali-soluble polysaccharides isolated from a *Boletus edulis* (Bull.) fruiting body and their potential for heavy metal biosorption. *Food Chem.* **2018**, *266*, 329–334. [CrossRef]
58. Kiho, T.; Yoshida, I.; Nagai, K.; Ukai, S.; Hara, C. (1→3)-α-D-glucan from an alkaline extract of *Agrocybe cylindracea*, and antitumor activity of its *O*-(carboxymethyl)ated derivatives. *Carbohydr. Res.* **1989**, *189*, 273–279. [CrossRef]
59. Lin, H.; Han, R.; Wu, W. Glucans and applications in drug delivery. *Carbohydr. Polym.* **2024**, *332*, 121904. [CrossRef]
60. Kagimura, F.Y.; da Cunha, M.A.A.; Barbosa, A.M.; Dekker, R.F.; Malfatti, C.R.M. Biological activities of derivatized D-glucans: A review. *Int. J. Biol. Macromol.* **2015**, *72*, 588–598. [CrossRef]
61. Venkatachalam, G.; Arumugam, S.; Doble, M. Industrial production and applications of α/β linear and branched glucans. *Indian Chem. Eng.* **2021**, *63*, 533–547. [CrossRef]

MDPI AG
Grosspeteranlage 5
4052 Basel
Switzerland
Tel.: +41 61 683 77 34

Foods Editorial Office
E-mail: foods@mdpi.com
www.mdpi.com/journal/foods

www.ingramcontent.com/pod-product-compliance
Lightning Source LLC
Chambersburg PA
CBHW062056220926
51850CB00010B/2783
* 9 7 8 3 7 2 5 8 5 9 8 1 8 *